Aluminum Alloys
for Transportation, Packaging, Aerospace and Other Applications

ALUMINUM ALLOYS FOR TRANSPORTATION, PACKAGING, AEROSPACE AND OTHER APPLICATIONS

TMS Member Price: $67 TMS Student Member Price: $54

List Price: $97

Related Titles

- *Shape Casting II: 2nd International Symposium,*
  edited by Paul N. Crepeau, Murat Tiryakioglu and John Campbell
- *Solidification of Aluminum Alloys,*
  edited by Men G. Chu, Douglas A. Granger and Qingyou Han
- *Trends in Materials and Manufacturing Technology and
  Powder Metallurgy R&D in the Transportation Industry:
  6th MPMD Global Innovations Symposium,*
  edited by Thomas R. Bieler, James W. Sears, John E. Carsley,
  Hamish L. Fraser and John E. Smugeresky

HOW TO ORDER PUBLICATIONS

For a complete listing of TMS publications, contact TMS at
(800) 759-4TMS or visit the TMS Document Center at http://doc.tms.org:

- Purchase publications conveniently online.
- View complete descriptions, tables of contents and sample pages.
- Find award-winning landmark papers and reissued out-of-print titles.
- Compile customized publications that meet your unique needs.

MEMBER DISCOUNTS

TMS members receive a 30% discount on TMS publications. In addition,
members receive a free subscription to the monthly technical journal *JOM*
(both in print and online), discounts on meeting registrations, and additional
online resources to name a few of the benefits. To begin saving immediately
on TMS publications, complete a membership application when placing your
order in the TMS Document Center at http://doc.tms.org or contact TMS.

Telephone: (724) 776-9000 / (800) 759-4TMS
E-mail: membership@tms.org or publications@tms.org
Web: www.tms.org

Aluminum Alloys
for Transportation, Packaging, Aerospace and Other Applications

Proceedings of a symposia sponsored by
the Light Metals Division of
TMS (The Minerals, Metals & Materials Society)

TMS 2007 Annual Meeting & Exhibition
Orlando, Florida, USA
February 25-March 1, 2007

Edited by

Dr. Subodh K. Das
Dr. Weimin Yin

A Publication of

A Publication of **The Minerals, Metals & Materials Society (TMS)**
184 Thorn Hill Road
Warrendale, Pennsylvania 15086-7528
(724) 776-9000

Visit the TMS Web site at
http://www.tms.org

Library of Congress Catalog Number 2007921346
ISBN Number 978-0-87339-662-2

If you are interested in purchasing a copy of this book, or if you would like to receive the latest TMS publications catalog, please telephone (724) 776-9000, ext. 270, or (800) 759-4TMS.

ALUMINUM ALLOYS FOR TRANSPORTATION, PACKAGING, AEROSPACE AND OTHER APPLICATIONS TABLE OF CONTENTS

Aluminum Alloys for Transportation, Packaging, Aerospace and Other Applications

Aluminum Applications

Aluminum Products

Alloy Development

Alloy Processing

Alloy Characterization

Alloys Mechanical Behavior

PREFACE

These proceedings of the symposium "Aluminum Alloys for Transportation, Packaging, Aerospace, and Other Applications" are a collection of contributed papers presented at the TMS 2007 Annual Meeting & Exhibition in Orlando, Florida, United States. The symposium is sponsored by the Light Metals Division.

The positive response to the symposium, from both academic institutes and the aluminum fabrication industry, indicates a continued strong interest in relating the basics of physical and mechanical metallurgy to the applied aspects of aluminum manufacturing. It is intended that this volume will continue to stimulate further research and development of aluminum alloys and their applications.

The editors wish to thank all authors and TMS staff members who made this symposium and the proceedings possible and successful.

January 16, 2007

Subodh K. Das
Weimin Yin

Secat Inc.
1505 Bull Lea Road
Lexington, KY 40511
USA

ABOUT THE EDITORS

Subodh K. Das is the president and chief executive officer of Secat Inc. and the director of the Center for Aluminum Technology, both based at the University of Kentucky, Lexington, Kentucky, United States. In this capacity, he directs research and development (R&D) activities for several aluminum companies. Dr. Das developed this unique R&D consortium of aluminum companies, universities, state and federal governments, and national laboratories. Among his other accomplishments, he received the prestigious Distinguished Aluminum Service award at The Minerals, Metals & Materials Society's (TMS) annual meeting in March 2000 and was elected Fellow by the American Society of Metals in 2002. Dr. Das served as chairman of TMS' Light Metals Division and on the board of directors. He also chaired the Technical Advisory Committee of the Aluminum Association based in Washington, D.C., United States. Dr. Das obtained his doctorate from the University of Michigan, Ann Arbor, Michigan, United States, in 1974 in metallurgical engineering.

Weimin Yin is a materials engineer at Secat Inc. He has been active for over a decade in the area of structured materials, including intermetallics, superalloys, nanostructured metals and aluminum alloys. Dr. Yin's research interest is in the advanced processing and characterization of aluminum alloys. He served a fellowship from the International Centre of Theoretical Physics, UNESCO, at an Italian national lab (CNR-TeMPE) from 1996 to 1997 and won the Noah A. Kahn ASTM Committee E-7 Award in 2001. Dr. Yin received his doctorate in materials science from Polytechnic University at Brooklyn, New York, United States, in 2003.

Aluminum Alloys

for Transportation, Packaging, Aerospace and Other Applications

Aluminum Applications

Aluminum Alloys for Transportation, Packaging, Aeropsace, and Other Applications
Edited by Subodh K. Das, Weimin Yin
TMS (The Minerals, Metals & Materials Society), 2007

CHARACTERIZATION OF SURFACE DEFECTS ENCOUNTERED IN TWIN ROLL CAST ALUMINUM STRIPS

Murat DUNDAR[1], Özgül KELES[1]

ASSAN Aluminum
E5 Karayolu 32. Km. Tuzla, İstanbul, 81700, Turkey

Keywords: Twin roll casting, Automotive alloys, Surface segregations

Abstract

Solidification mechanism in twin roll casting (TRC) technique provides various advantages for the production of aluminum alloys having narrow solidification range. High production volume in casting, reduced cost of rolling due to the gauge of initial material to be rolled and elimination of some other intermediate processes are among those. However, alloys having wide solidification range is in tendency of creating serious surface defects.
They can readily deteriorate mechanical performance and results in severe complications not only in demanding applications, but also in ordinary engineering applications. If exposed to a surface treatment, they can be easily revealed by resulting in impairment of the aesthetic appearance. Present study aims microstructural characterization of surface defects created in TRC technique and their affects on some critical applications. 5000 and 3000 series aluminum alloys produced for demanding applications and 8000 series alloys for packaging industry form the subject of this study.

Introduction

Competitive environment in aluminum industry forces to investigate alternative solutions for existing technology in various ways. Recent studies on twin roll cast aluminum strips, including 5000 series alloys, provided promising results for applications necessitating major cost reductions in materials issue[1-4]. As compared to the traditional hot mill process, the relatively low capital cost of twin-roll casters, in combination with their lower energy and labor cost, have made twin-roll casting an increasingly popular method of producing a wide range of aluminum flat rolled products. The most attractive aspect of twin-roll casting is the ability to cast at thinner gauges allowing significant reductions in the cost of further down stream operations, along with higher productivity in the casting process itself.
The advantages inherent in the twin-roll casting process do not in itself ensure a competitive advantage in the production of strip if the strip is not of commensurable price and acceptable surface quality. Due to the nature of very high solidification rates achieved in thin strip casting, microstructural components of the as-cast strip exhibit unique features as compared to their DC cast and hot rolled counterparts. Much finer and uniformly distributed intermetallic constituents of twin roll cast strips enhance mechanical performance of the material for stringent applications. Topographical and microstructural constituent of the strip surface are determined by the characteristics of the solidification. Any effect distracting the solidification path of the metal,

after the deployment from the ceramic caster nozzle, leads to local microstructural abnormality compared to its vicinity. Free surface of the strip, that is the interface between caster roll surface and the melt, is more prone for those defects to be observed. Since outermost skin of the strip is not removed by any mechanical means, contrary to the scalping operation of DC cast and hot rolled ingots, through thickness uniformity in microstructural features is an important requirement for the performance of material in different applications, including aesthetic properties. The severity of surface segregations was found to depend critically on alloy composition [5].

Surface defects of twin roll cast strips originating from the casting phase of production were investigated in the present study. Twin roll cast aluminum alloys employed for general engineering applications, packaging and automotive applications were characterized for their surface defects encountered.

Experimental

Surface defects of various samples cast at 5-6 mm and samples gathered from rolled strips out of these coils were the subject of present study. 2200 mm wide strips were produced by employing industrial size twin roll casters having roll diameter between 1060-1100 mm. While some type of defects are readily detected at the as-cast surface of the strips, further rolling operations that decrease the surface roughness help others to be detected by visual inspection at thinner gauges. Optical microscope (Olympus PME3) and SEM (JEOL5600) equipped with EDS unit (Oxford 6587) were used for microstructural characterization of the defects. Surface roughness of as-cast strip and rolled material does not allow microstructural constituents to be detected unless any prior preparation technique is applied. Therefore, samples were slightly polished with 0,3 μm SiO_2 for those constituents to be revealed for analysis.

Severe surface segregations can have detrimental effects on mechanical performance of the material. If there exist any severe bending or deep drawing operation involved in the application, these defects lead to premature failure. Therefore, cross section of samples was polished by using metallographic sample preparation techniques to create very sharp corners between the plane of cast strip and thickness direction. The samples prepared with this method were bent parallel to the rolling direction to elucidate interaction between the tensile stress field at the convex side of the sheet and the constituents of surface segregations. Chemical content of intermetallic particles forming clusters were analyzed with the help of semi-quantitative EDS technique.

Results

The Al-Mg alloys (5000 series) are extensively used in applications requiring a good combination of strength and formability. Typical example of this series is AA5754 based on Al-3wt%Mg, which is extensively used in automotive structures and engineering applications. Due to its wide solidification temperature range and high oxidation tendency in liquid form, transfer of metal from furnace to the caster and delivery through the ceramic nozzle bears some potential problems, if some critical rules are violated in casting practice of TRC technique. Physical solidification path of the molten metal prior to the contact with the caster rolls has strong influence on the solidification abnormalities. Disruption of the liquid or semi-solid metal by any physical mean results in formation of surface segregations. Surface characteristics of caster rolls can introduce supplementary effects on the formation mechanism of surface segregations. These are closely related to the "wetting" of the liquid metal on roll surface. Figure 1 shows heavy surface segregations on the as-cast surface of AA5754. Particles were revealed by slightly polishing the surface to be easily observed with optical microscope.

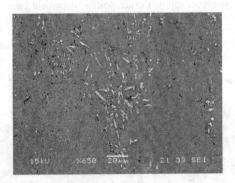

Figure 1. Surface segregations on the as-cast AA5754.

Wide solidification range of an alloy makes it more prone to form microstructural defects in the casting process, unless casting parameters are optimized. As the alloys that can be produced with TRC is listed in the descending order of their solidification range, while AA 5182 has the widest range, AA1050 has the narrowest solidification range with the lowest surface segregation tendency.

SEM and EDS studies carried out to determine chemical content of individual particles reveal that intermetallic particles are majorly composed of Al-Fe, Al-Fe-Si, Al-Fe-Mn and Al-Fe-Mg, depending on the alloy composition. They are relatively hard and non-deformable particles compared to the surrounding ductile matrix. While they can be formed within a narrow but continuous band along the casting direction, whole surface can also be covered with those if casting parameters are not appropriately controlled. Figure 2 shows one of these bands after slightly etching. Especially, in the latter case, the strain hardening and fracture mechanism of limited volume containing intermetallic particles are analogues to the mechanism operating in particulated reinforced aluminum alloys [6,7].

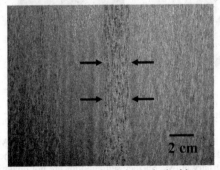

Figure 2. Segregation band, after etching, is marked with arrows.

Twin roll cast aluminum strips are cold rolled to the required thickness in their processing route. Due to the frictional forces at the interface between the work rolls of the mill and the sheet, very large shear stresses are generated [8-10]. On the other hand, deformation characteristics of rolling, that is plane strain, help alignment of these particles parallel to the rolling direction. These particles have relatively large aspect ratios. Corner profile of particle causes intense plastic straining in the nearby matrix and leads to void nucleation at the corners (Figure 3)[11-12]. This mechanism is more pronounced if there are clustered particles within a narrow domain. Overlapping strain fields around the corners of clustered particles result in coalescence of voids with induced more strain during rolling.

(a)   (b)

Figure 3. Void nucleation at the corner of intermetallic particles (a) and expanded voids due to coalescence of strain fields (b).

Figure 4 shows the typical giant intermetallic particles formed during casting of AA5754. Al-Fe-Mg or Al-Fe are the major intermetallic phases. Dot mapping of a marked area is given in Figure 4. The chemical content of these phases are determined by the alloy composition.

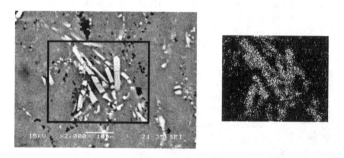

Figure 4. Dot mapping of intermetallic phases in AA5754.

AA8011was another alloy on which surface segregations were investigated. Major alloying elements of AA8011 are Fe and Si that form Al-Fe-Si and Al-Fe phases. They are typical intermetallic phases of Al-Fe-Si ternary alloys. Surface segregations are not unavoidable even in such a low alloyed materials. Unless casting related sources are eliminated, surface segregations can readily be generated at the sheet surface. Due to its lower solidification range, AA8011 is less prone to form surface segregations, compared to that of AA5754 involving more than 2,6% Mg and other alloying elements, like Fe and Si. However, contrary to the AA5754, morphology of the intermetallic phases are significantly different than that of AA5754. They are in equiaxed shape rather than whisker-like geometry. Their aspect ratios are close to one. Cold rolled samples have shown that equiaxed morphology of particles is not even sufficient enough to prevent void nucleation after applying several rolling passes (Figure 5). Severe void formation is again observed around the intermetallic clusters.

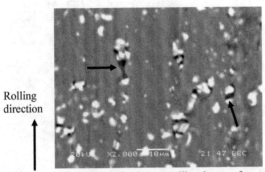

Rolling direction

Figure 5. Void formation around the intermetallic phases after several passes. Note rolling direction.

Elemental analysis of the individual particles, formed in AA8011, was conducted with EDS technique. It revealed that they were Al-Fe-Si based phases (Figure 6). Although their chemical compositions are determined with semi-quantitative technique, their exact stochiometry, crystallographic nature and growth characteristics require further investigation.

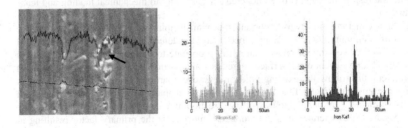

Figure 6. EDS analysis of intermetallic phases in AA8011, after rolling. An excessively opened void is marked with an arrow.

Initial form of these surface segregations was investigated on the as-cast strip surface. By employing metallographic techniques, as-cast strip was slightly polished. Clustered intermetallic phases were detected at the same position where segregation band was spotted on the rolled strip. As cast strip surface and rolled surface are shown in Figure 7 (a) and (b), respectively. Figure 7 (b) illustrates the noticeable border between segregated (left side) and segregation free areas.

(a)   **(b)**

Figure 7. Clustered segregations on the as-cast strip surface (a) and after rolling (b).

Surface segregations can be very detrimental for many end-use of aluminum sheet and foils products. As the gauge of the strip decreases with rolling passes, their volume fraction increases through thickness. Some applications, in which plane strain deformation is operative at particular areas of their geometry, such as deep drawing, these non-deformable particles may lead to premature failure. Analogues to the Marciniack-Kuczinsky principle[13], any physical defects, either in the form of local thinning through the thickness or microstructural discontinuity compared to the rest of the volume or plane, create a sensitivity to mechanical loading and leads to premature failure.

Hemming and clinching are two important mechanical joining techniques in automotive applications of aluminum alloys. In the presence of surface defects, even with the application of large bending radius, very minor cracks can easily propagate to the inner side of the thickness with the help of above mentioned surface segregates and associated voids. Crack propagation pattern preferentially follow the existing voids and cracked particles. Figure 8 illustrates crack propagation pattern at the outer fibers of an AA5754 sheet that involve segregates. The sheet thickness is 1 mm and bent parallel to the rolling direction, that comply the rules specified in the standards, Shear stress generated at sheet-work roll interface is the primary factor resulting in fragmentation of these non-deformable particles, having large aspect ratio. As, these intrinsically existing defects are exposed to the tensile stresses at the convex side of the bending radius, they form an easy path for the propagating crack.

Detrimental effects of these defects are not limited with the mechanical performance of the material. Surface segregates encountered in the sheet of AA8011, aimed to be employed for production of lid foil, easily alter surface roughness that also leads to pronounced effect on the lacquering performance of the surface. The bands that bear surface segregations develop different surface roughness compared to the adjacent areas, due to voids around individual or clustered segregates. Since wetting characteristics of these areas are significantly different than that of adjacent segregate-free sites, they hold more lacquer and these are visually recognized as dark lacquered areas or shades.

Figure 8. Crack propagation pattern at the outer fibers of AA5754. Free surface is exposed to the tensile stress in bending.

Conclusions

1. Aluminum alloys having wide solidification range is prone to develop surface segregations during Twin roll casting.
2. Casting parameters have significant influences on the formation mechanism of surface segregations, size and distribution on the as-cast strip surface. Surface segregations can be controlled and eliminated with optimized casting parameters.
3. They can not be dissolved or eliminated by any means of post metallurgical process.
4. Excessive shear stresses generated at the outer fibers of sheet during cold rolling cause these hard and non-deformable particles to be broken into small fragments.
5. Fracture mechanics and crack propagation mechanism is dictated by those particles and mechanical performance of the strip is significantly deteriorated due to the presence of these particles. They allow limited bending, hemming and forming applications.
6. Presence of surface segregations has adverse affect if the strip is exposed to an electrochemical process or coating.

References

1. Birol, Y., Kara, G., Dündar, M., Akkurt, A.S. and Romanowski, C., **2001**, TMS, *Automotive Alloys*, ed. Das, S., Pa.
2. Dündar, M., Birol, Y., Slamova, M., Akkurt, A.S., Romanowski, C., **2001**, *Materials Week*, Munich.
3. Dündar, M., Sarıoğlu, A.S. and Akkurt, A.S., TMS **2002**, ed. Das, S.,Pa.
4. Dündar, M., Birol, Y., Akkurt, A.S., **2002**, *ICAA8*, Cambridge, UK.

5. Lockyer, S.A., Yun, M., Hunt, J.D., and Edmonds, D.V., Materials Characterization 37: 301-310 (1996), "Micro and Macrodefects in thin sheet Twi-roll cast Aluminum alloys".
6. Humphreys, F.J. and Kalu, P.N., 1990, Acta Metallurgica, "The Plasticity of Particle Containing Polycrystals", 6, 917-930.
7. Lewandowski, J.J. and Liu, C., 1989, Materials Science and Engineering, "Effects of Matrix Microstructure amd Particle Distribution on Fracture of an Aluminum Metal Mtarix Composites", A107, 241-255.
8. Truskowski, W., Krol, J. and Major, B., 1980, Metallurgical Transaction, "Inhomogenity of Rolling Texture in FCC Metals", 11A, 749.
9. Asbeck, H.O., and Mecking, H., 1978, Materials Science and Engineering, "Influence of Friction and Geometry of Deformation on Texture Inhomogenities During Rolling of Cu Single Crystals.
10. Dundar, M. 2001, Assan Aluminum Report UPG/CRN/026.
11. Nutt, S.R. and Needlemman, A., 1987, Scripta Metallurgica, "Void Nucleation at the Fiber Ends in Al-SiC Composites", Vol. 21, 705-710.
12. Nutt, S.R. and Duva J.M., 1986, Scripta Metallurgica, "A Failure Mechanism in Al-SiC Composite", Vol. 20, 1055-1058.
13. Marciniak Z. and Kuczynski, K. 1967, *Int. J. Mech. Sci.*, 9, p. 609.

Aluminum Alloys for Transportation, Packaging, Aeropsace, and Other Applications
Edited by Subodh K. Das, Weimin Yin
TMS (The Minerals, Metals & Materials Society), 2007

Life-cycle Cost Analysis:
Aluminum versus Steel in Passenger Cars

C.A. Ungureanu[1], S. Das[2], I.S. Jawahir[1]

[1]University of Kentucky, Lexington, KY, 40506, USA
[2]Secat Inc., 1505 Bull Lea Road, Lexington, KY, 40511, USA

Keywords: Aluminum, life cycle, weight reduction, recycling.

Abstract

In light of escalating fuel p rices and the ongoing clim ate cha nge discussion, sustainability considerations are currently tak ing a m ore prom inent role in m aterial selec tion d ecisions f or automotive applications. This paper presents a new methodology for total life-cycle cost analysis and employs a case study involving the use of al uminum in autom otive applications. This study is aimed at developing a new su stainability model to quantify the total cost encountered over the entire life-cycle of a vehicle considering all four life-cycle stages: (1) pre-m anufacturing, (2) manufacturing, (3) use and (4) pos t-use. Also, the paper presents a quantitative evaluation of the environmental im pact of using aluminum m aterial in a veh icle. The p aper com pares the use of aluminum with the trad itional use of steel alloys in a given autom otive application by providing details of economic and environmental performance of the vehicle over the total life-cycle.

Introduction

Reducing manufacturing costs a nd tailpipe em issions by us ing light-w eight m aterials which can easily be recycled or reused are am ong the m ajor issues in today's automobile industry. The growing emphasis on total cost and environm ental im pact has forced the life-cycle co st issue to b e the driving factor. Weight saving s in the ove rall car m ass is considered to be a m ajor research f ocus. Alum inum is proven to be among the potentia l materials capable of achieving weight reduction without sacrificing the vehicle safety and performance. Despite significant technological advantages in aluminum alloys, the use of aluminum alloys to repla ce traditional materials such as steels has been slow due to lack of comprehensive economic analysis of the entire life-cycle of automotive products.
In considering the total life-cycle of an autom obile covering four stages (pre-manufacturing, m anufacturing, use, and post- use), it is apparent that during the operational (use) stage of a vehicle, alum inum is proven to be a reliable alte rnative for traditional materials currently used in autom otive body stru ctures largely due to its cost-effectiveness and superior perform ance due to light weight. W ith the ga s price variation, the initial cost advantage of using steel in body components gained in pre-m anufacturing and m anufacturing stag es can b e o vercome during th e op erational (u se) s tage of the vehicle, sin ce the ligh ter alternative provi des significant savings in term s of f uel consumption and consequently generation of ai rborne gas em issions. Also, the superior recyclability and reus ability of aluminum in the post-use stage outweighs the trad itional materials despite the higher cost involved in producing primary aluminum.

This paper presents a system atic study of the total life-cycle cost analysis and the environmental impact of using aluminum-based automotive products. This study is aimed at developing a new model to quantify the total cost encountered over the entire life-cycle of a vehicle considering material su bstitution in the body structure of the vehicle, since the so-called body-in-white (BIW) structure plus exterior cl osure panels represent an important group where significant weight savings can be achieved. Also, the environmental im pact over the lifetim e of the vehicle is being quantified. Overall, the study concludes that considering the entire life-cycle of an automobile, from extraction of materials to the f inal d isposal including recycling and re use applications, alum inum proves to be a potential alternative for steels in future automotive applications.

Major Assumptions

Knowing that the greatest opportunity for we ight savings com es from the body structure and exterior closure panels, and that a dditional weight reduction can be achieved by downsizing the other components such as e ngine components [1, 2], the proposed model considers achieving weight reduction by repl acing the conventional m aterial used in vehicle's construction (i.e., steel) with a lighter mass equivalent material (i.e., aluminum), maintaining the sam e vehicle design and us ing the sam e m anufacturing processes for body components. The major assumptions for this study are listed in Table 1.
The starting value for gas price is assum ed to be $2.30 per gallon, a value which is considered to be closer to the cu rrent gas price. The gas p rice can flu ctuate, and a 20 percent increase or decrease for th e curren t value has been considered in the current study. Thus, the resulting price range is be tween $1.84 and $2.76 per gallon as shown in Table 1. For the pre-m anufacturing stage, th e cost calculations for both m aterials are based on the assumption that 308 kg of aluminum sheet would be required to produce the completed 193 kg aluminum body structure and 565 kg steel sheet are needed to produce 371 kg steel body structure. According to Stodols ky [1], the prim ary material used in the typical passenger car today is steel, which can be purchased for a cost between $0.77 and $1.20/kg. A 20 percent increase or decrease for st eel sheet cost has also been considered, with a range of values between $0.63 – $1.17/kg. Since aluminum is a m aterial which is likely to replace steel in autom otive body components [3], the starting value for aluminum s heet has been chosen as $3.3/kg [1]. A 20 percent increase or decrease in aluminum sheet cost has also been consider ed, giving a range of values between $2.31 - $4.29/kg. The starting values for both m aterials are considered to be in agreem ent with the generally known fa ct that the cost to produce prim ary al uminum is between 2 to 5 times more expensive than the cost to produce primary steel [4, 5].
For the m anufacturing stage of the life-cycl e, the calculations use Technical Cost Modeling software developed at MIT [3, 6] for a production volum e of 150,000 vehicles per year. The analysis considers both fabrication costs and assembly costs encountered by the body-in-white (BIW) structure and the exterior panels during the m anufacturing stage. The f uel consum ption of vehicles is assum ed to be constant throughout the use stage, with a lower vehicle weight providing im proved fuel efficiency. It is assum ed that 5 % fuel efficien cy can be ach ieved from a 10 % weight-reduction [3, 5]. In the case of steel BIW, the fuel economy has been assumed to be 20 mpg, whereas the fuel efficiency for aluminum BIW is assumed to be 22 mpg [2].

Table 1: The basic assumptions of major parameters used in the current study

Parameter	Starting value	Range
Gas Price ($/gal)	2.30	1.84 – 2.76
Cost of Steel ($/kg)	0.90	0.63 – 1.17
Cost of Aluminum ($/kg)	3.30	2.31 – 4.29
Price of Scrap ($/kg) Steel Aluminum	0.09 0.93	0.069 – 0.129 0.657 – 1.221
Fuel Consumption (mpg) Steel BIW Aluminum BIW	20 22	
Total Vehicle Weight (kg) Steel BIW Aluminum BIW	1,418 1,155	
Body-in-White Weight (kg) Steel Aluminum	371 193	
Life of the Car (years)	14	
Miles Driven in Year 1	15,220	
Lifetime Miles Driven	174,140	
Recycling Percentage Steel Aluminum	90 90	

The life time of the veh icle has b een assum ed 14 years [7]. The total num ber of m iles driven over the life ti me of the vehicle is 174,140 m iles, with the assumpti on that in the first year, the vehicle is driven 15,220 m iles, and that the num ber of m iles driven annually decreases as th e vehicle age increases as shown in Table 2. The price values of scrap material and recycled material are listed in Table 3 for both materials [8].

Once the vehicle reaches its end-of-life, it is considered that the owner sells the vehicle to a dism antler and that 90 percen t of the BIW m aterial is r ecycled [9, 10]. It is als o considered closed-loop recy cling of obsolete autom otive BIW materia ls, where the recycled materials are returned to their original usage through further processing.

Table 2: Estimated annual miles driven by the vehicle age

Vehicle Age (Years)	Annual Miles Driven	Total Miles Driven

1	15,220	15,220
2-5	14,250	72,220
6-10	12,560	135,020
10-14	9,780	174,140

Table 3: Material price database for aluminum and steel

Material	Price ($/kg)	Scrap ($/kg)	Recycle ($/kg)
Steel	0.9	0.09	0.12
Aluminum	3.3	0.93	1.32

Apart from the cost analysis, the model also quantifies the am ounts of carbon dioxide emissions generated during th e processing of the m aterials, m anufacturing the body structures, use of the vehicle, and in recycling the materials. For all four life-cycle stages, carbon dioxide em issions for both m aterials are listed in Table 4 and these values are derived from [11]. The current m odel trac ks only carbon dioxide em issions associated with fuels used for alu minum and steel opera tions during each stag e. Other fuel-related emissions such as carbon m onoxide, nitrous oxides, sulfur dioxide, and other com pounds are not considered in this study.

Table 4: Total carbon dioxide emissions for steel and aluminum BIW (Year 1)

Stage	Steel (kg CO_2/BIW)	Aluminum (kg CO_2/BIW)
Pre-manufacturing	1,913.5	2,689
Manufacturing	19.5	18.6
Use	6,772.5	6,139.5
Post-use	282.5	75.7

Being a highly energy-intensive process, producing virgin alum inum generates more carbon dioxide em issions than producing vi rgin steel. Since their m anufacture and assembly processes are assum ed to be sim ilar, the am ounts of carbon dioxide generated during the m anufacturing stage di ffer slightly, being the direct re sult of using e lectricity to operate the m achinery. The v ehicle's op erational (u se) stag e h as the greatest environmental im pact in term s of carbon di oxide em issions. Fuel economy, the number of years the vehicle is used on the roads and the em issions rate are am ong the most common fa ctors contributing to the am ount of carbon dioxide generated over the operational stage. The lighter alternative is proven to emit less gaseous substances since it needs less power to m ove and therefore less fuel. Credits for em ission rates are given in accordance with the U. S. Environmental Protection Agency recommendations [12]. For the post-use stage, the amounts of carbon diox ide generated by both materials, are based on the assumption that 90 percent o f the material is recycled once the vehicle reach es its end-of-life [9] and that the recycled alum inum saves 95 percent of the energy to produce virgin aluminum [13, 14] whereas the recycled steel saves between 40-75 percent of the energy requ ired to prod uce virg in steel [10]. All the abov e values ar e illus trative, no t definitive and they are derived fro m published sources which helped in developin g the model. By changing the starting values a ccording to the actual cons ent and realistic estimates, the model will recalculate all the costs encountered by the BIW structures over the entire life-cycle of the vehicle.

Preliminary Results

Fuel economy, gas price variation and the   number of m iles driven on the roads are important parameters which make up for the total cost encountered by the vehicle during the use stage. The cost of gasoline encounter ed over the operationa 1 (use) stage of the vehicle is a function of the ga s price variation, for both m aterial scenarios, and is shown in Figure 1. As expecte d, alum inum substitution would pr ovide im portant s avings over the entire range of the gas price variation. At a price of only $2.30 per gallon and a fuel economy improvement of 10 percent, it is shown that over the life time of the vehicle (14 years), approximately 791.5 gallons of gasoline can be saved. This number translates into about $1,820 saved over the same period of time.

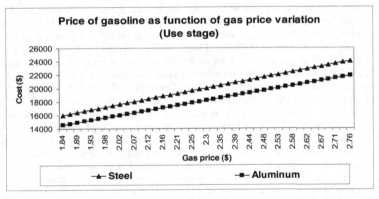

Figure 1: Cost of gasoline as a function of gas price variation (Use stage, 14 years)

The carbon dioxide em issions for the "Use" stage depend on the num ber of miles driven, fuel economy, and the em issions rate. Acco rding to the U S Environmental P rotection Agency, it is assum ed 0.916 pounds of CO_2 emissions per mile for a passenger car's fuel consumption of 21.5 m iles per gallon. Sin ce carbon dioxide em issions are directly proportional to fuel economy, each 1% increase (decrease) in fuel consumption resu lts in a corresponding 1% increase (decrease) in car bon dioxide em issions [12].Therefore, this study considers for alum inum BI W structured vehicle, 0.88 pounds CO_2 e missions per mile and for steel BIW structured vehicle 0.98 pounds CO_2 emissions per mile. The CO_2 emissions generated during the use stage as function of the number of years are shown in Figure 2.

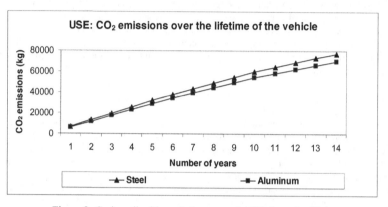

Figure 2: Carbon dioxide emissions over the lifetime of vehicle

Since the co st encountered during the "Use" stage has the highest im pact on com puting the total ow nership cost and the number of m iles driven, the recycling conten t and the price of gas are important parameters to compute the total cost encountered by the vehicle over its life-cycle. This study com pares the to tal costs encountered by vehicle for three different m ileage scenarios (15,220 m iles, 57,970 m iles, and 135,020 m iles). Four different levels of recycled material, for each m ileage case scenario, are also cons idered: 0, 25, 75, 100 percent, both recycl ed materials (steel and alum inum), and a special case scenario, in which 75 percent alum inum and 25 percen t steel is recy cled m aterial. Pre- manufacturing costs depend greatly on the pe rcent of m aterial r ecycled. W ith the increased use of recycled material, the material cost becomes smaller. The manufacturing costs consider both the cost of body fabricati on and the cost of final assem bly. The cost functions for alum inum and steel sheets a nd the fabrication costs for body com ponents differ, and it is shown that steel fabrication cost is less th an the fabrication co st for aluminum body com ponents. Since the assem bly cost for alum inum body structure is higher than the assembly cost for steel body structure, the manufacturing costs to produce steel body structure are genera lly lower than the m anufacturing costs to produce the aluminum body structure. Cost s encountered during the "Use" stage of the vehicle are functions of the num ber of m iles driven, fu el consum ption, and pr ice of gasoline. An improvement in fuel consum ption, and the incr ease in the num ber of m iles driven by the vehicle lead to an increase in the difference between the number of gallons of gas used by the s teel s tructured veh icle and the num ber of gallons of gas used by the alum inum structured vehicle, thus, m aking alu minum BI W vehicle much cheaper in term s of the money spent on gasoline during this stage. The "Post-use" stage costs consider only obsolete scrap from the end-of-life vehicle. Si nce both m aterials are con sidered to b e 90 percent recycled, and that alum   inum has   a higher scrap value, $0.94 per kilogra   m compared to $0.10 per kilogram   for steel,   alum inum has a higher post-use value. Figure 3 refers to the first m ileage case s cenario (15,220 m iles driven) for Year 1, and it shows the ratio of the total cost for aluminum versus the total cost for steel over the entire life-cycle of the vehicle as function of gas price variation. As content of material recycled is increased, for instance from 25 % to 75 % material recycled, the ratio becomes closer to the unity value, bu t still the total cost f or steel BIW is smalle r than the total cost f or aluminum BIW for the entire range of gas   price variation. Howeve r, a 100% recycled material us for both m   aterials would   give a cost advantage for aluminum.

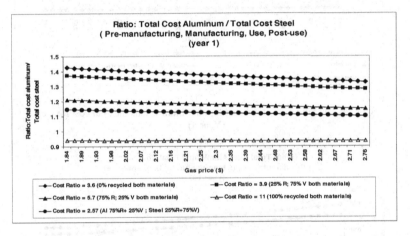

Figure 3: The ratio of the total cost for aluminum versus the total cost for steel
(Year 1)

Figures 4 and 5 show the sam   e decreasing tre nd for all s cenarios of recycled m aterial content, but for different num ber of m iles driven: 57,970 miles (Figure 4) and 135,020 miles (Figure 5), driven at Year 4 and Y ear 10, respectively. The difference between the total costs for alum inum and the total costs for steel reduces, as the difference between the "Use" stage costs becomes larger. After 135,020 miles driven (Year 10), the total cost ratio is less than the unity value, for alm   ost all scenario s of recycled m aterial con tent. Considering the case scenario where aluminum 75 percent and steel 25 percent m aterial recycled, Figure 6 shows the total ownership   cost breakdown for both m   aterials.

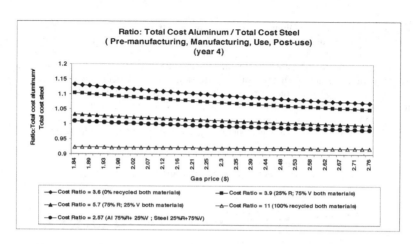

Figure 4: The ratio of the total cost for aluminum versus the total cost for steel
(Year 4)

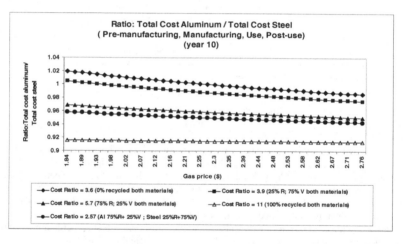

Figure 5: The ratio of the total cost for aluminum versus the total cost for steel
(Year 10)

Being a cheaper m aterial to produce and m anufacture, for the first f our years of vehicle
usage, steel BI W structure is shown to be a more economical option. Once the vehicle's
usage is in creased, the difference between the use costs f or both m aterials beco mes
significant, making alum inum BI W structure a more econom ical option. After ten years,
the aluminum structure has a cost advantage of about 5 percent over the steel structure.

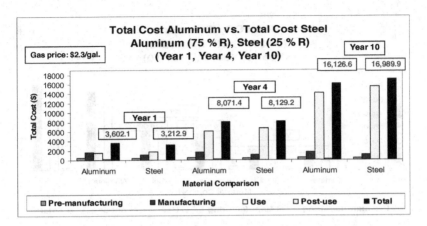

Figure 6: Total cost breakdown (Aluminum vs. Steel) for all four life-cycle stages

For the pre-m anufacturing stag e, the am ount of carbon dioxid e generated is calculated based on the content of m aterial recycled. Figure 7 shows the am ounts of carbon dioxide generated during this stage for increasing recycling rate for both materials.

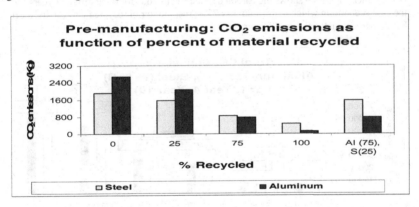

Figure 7: Carbon dioxide emissions as a function of recycled material content during the pre-manufacturing stage

For the m anufacturing stage, the amounts of carbon dioxide em issions are quite sim ilar (19.5 kg CO_2 emissions for manufacturing al uminum BIW structure and 18.6 kg CO_2 emissions for m anufacturing steel BIW structure) while the m anufacturing processes are assumed to be different.

Figure 8 shows the carbon dioxide em issions in all four life-cycl e stages, for three different years, for the case of usi ng zero percen t recycled m aterials.

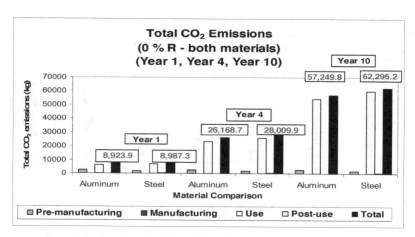

Figure 8: Total carbon dioxide emissions breakdown (0 % R both materials)

Even though the production of virgin alum inum is highly energy-inte nsive, it takes only one year of vehicle usage for alum inum to offset the carbon dioxide em ission disadvantage from the pre-m anufacturing st age, as a result of fuel consum ption improvement. Figure 9 shows the carbon dioxide emissions for three different years for the case scenario in which alum inum has 75 pe rcent material recycl ed content and steel has 25 percent material recycled content.

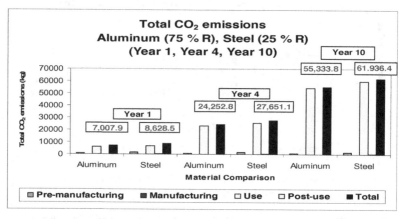

Figure 9: Total carbon dioxide emissions breakdown (Al. 75 % R; steel 25 % R

Fuel efficiency and energy savings from the use of recycled materials reduce dramatically the total amount of carbon dioxide generated by aluminum BIW structure over the entire life-cycle. The carbon dioxide emissions for aluminum BIW structure are about 8 percent lower than those for steel BIW structure after only one year of vehicle usage.

Summary and Future work

This study considers material-substitution as a m eans to ac hieve weight reduction, and the shows its benefits by considering the entire life-cycle of the vehicle, from fabrication of raw m aterials to th e f inal disposal. This work highli ghts the advantage of using aluminum i n auto body structures, from both e conomical and environm ental points of view by using a case study at a single-product level. Reducing the weight of the vehicle has a significant effect upon its lifetime monetary cost, since the cost at the "Use" stage presently co nstitutes a d ominant por tion of the overall cost. As the real gasoline pr ice increases and vehicle life is extended, the light weight issue beco mes even more important. Previous research has dem onstrated the cost advantage of producing automotive com ponents from virgin steel. The other two stages (use and post-use) were not considered significant for com puting the total life-cycle cost, since the gas price was considered to be low and recycling facilities for metals were not very well developed [3]. Considering zero percent recy cled content bo th m aterials, the in itial f abrication and manufacturing cost advantage fo r steel structure is offset by the lower costs for gasoline, and the higher m etal scrap value for alum inum structure in the use and post-use stages. This model shows that it takes 9 years or 122,460 miles, at a gas price of $2.53 per gallon for aluminum structured vehicle to offset the total cost for steel structured vehicle. As the gas price increases, at a valu e of $2.76, the total cost for aluminum structured vehicle ($18,355) becom es lower than the total cost for steel structur ed vehicle ($18,490). Furthermore, increasing the content of material recycled to 25 percent for both materials, the num ber of years the alum inum BI W needs to offset the total costs encountered by steel BI W drops to 7. It is shown that af ter 97,340 m iles, at a gas price of $2.76 per gallon, aluminum structured vehicle offsets the total cost of st eel structured vehicle. For 75 percent both m aterial recycled, it takes only 4 years or 57,970 m iles at a gas price of $2.66 for alum inum structure to of fset the tota l cost for steel structure. Under the most likely case scenario, (alum inum 75 percent and steel 25 percen t recy cled), th e model shows that after 3 years or 43,720 m iles at a gas price of $2.76 per gallon, alum inum BIW structure offsets the total costs of steel BIW structure as shown in Table 5.

Table 5: Total cost breakdown for (aluminum 75%, steel 25 % material recycled)

Stage	Aluminum cost ($)	Steel Cost ($)
Pre-manufacturing	559.3	398.4
Manufacturing	1,614.8	1,097.5
Use	5,484.8	6,033.3
Post-use	163.2	33.8
Total Cost	7,495.7	7,496.05

Figure 10 shows the total ownership cost breakdown encountered by both materials during each stage, after three years, at a gas price of $ 2.76.

21

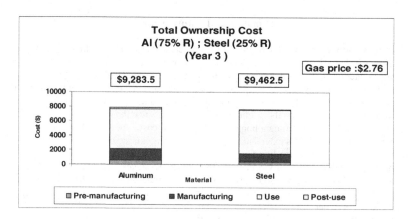

Figure 10: Total ownership cost

Regarding carbon dioxide em issions, the m odel shows the benefit of using lighter materials in the body construction of vehicl es. Figure 11 illu strates the to tal carbon dioxide emissions, over the vehicle's life-cycle considering that both are virgin m aterials. Despite the emission disadvantage from the pre-manufacturing stage, it is found that only one year or 15,220 m ile driven, needs for al uminum BI W structure to em it less carbon dioxide than the steel counterpart.

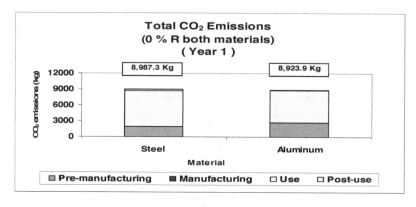

Figure 11: Total CO_2 emissions (both are virgin materials)

The energy savings from the recycled steel are not as dramatic as the energy savings from the recycled alum inum. The amount of carbon dioxide generated in producing the steel sheet with increas ed content of m aterial recy cled is not so drastically low, as that of the amount of carbon dioxide generated in produc ing the alum inum sheet with increased content of recycled m aterial. Using increased content of alum inum recycled m aterial in the vehicle's body, whi ch dram atically reduces the am ount of carbon dioxide generated in the process of m aking virgin alum inum, aluminum BIW structure is proven to em it about 7 percent less carbon dioxide than what steel BIW structure does em it, after only one year of vehicle usage. As the vehicle continues to "age", th e carbon dioxide savings increase, an d after ten years, there will be about 11 percent carbon dioxide em issions savings from the use of recycled aluminum in the vehicle's body structure (Figure 12).

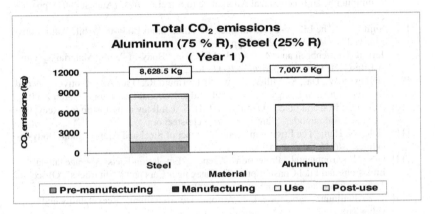

Figure 12: Total carbon dioxide emissions (Aluminum 75 % R, Steel 25 % R)

Based on these findings, and from the econom ical and environm ental benefits of using both m aterials, future work should be focu sed on determ ining the right com bination of these two materials in autom otive industry. This would help to reduce total costs and greenhouse gas emissions over the life-cycle of the vehicle and to im prove the safety and performance. Since take-back options are fast becom ing a n inevitable and unavoidable for car m akers, it would be essential to qua ntify and estim ate the total life-cycle cost encountered by the vehicles by considering tw o options: reuse of pa rts, and the use of recycled materials.

References:

[1]   Stodolsky, F., A. Vyas , R. Cuenca, a nd L. Gaines, "Life-Cycle Energy Savings Potential f rom Alum inum-Intensive Vehic les", *SAE 1995 Total Life Cycle Conference and Exposition*, Vienna, Austria (October 16 – 19, 1995).

[2]   Linda Gain es and Roy Cuenca, "Operation o f an Alum inum-Intensive Vehicle: Report on a Six-Year Project", Center for Transportation Research, Argonne National Laboratory, 1994.

[3]   Helen N. Han, Joel P. Clark, "Lifctim e Costing of the Body-in-W hites: Steel vs. Aluminum", *JOM*, May 1995.

[4]   Fridlyander, I.N., V.G. Sister, O.E. Grushko, V.V. Berstenev, L.M. Sheveleva and L.A. Ivanova, "Alum   inum Alloys: Prom   ising Materials in the Autom   otive Industry", *Metal Science and Heat Treatment*, September 2002, pp. 3-9.

[5]   Frank Field, Randolph Kirchain and Jo   el P. Clark, "L   ife-Cycle Assessm ent and Temporal Distributions of Em   issions", *Journal of Industrial Ecology*, Volum e 4, Number 2, 2001.

[6]   Anish Kelkar, Richard Roth, and Jo   el P. Clark, "Autom   otive Bodies: Can Aluminum be an E conomical Alternative to S teel?", *JOM* (August 2001), pp. 28-32.

[7]   Sujit Das, "The Lif   e-Cycle Im pacts of Alum inum Body-in-W hite Autom otive Material", *JOM* (August 2000), pp. 41-44.

[8]   Jeff R. Dieffenbach an d Anthony E. Mascarin, "Body-in -White Material System s: A Life-Cycle Cost Comparison", *JOM* , June 1993.

[9]   Hadley, S.W., Das, S., Miller, J.   W., "Alum inum R&D for Autom otive Uses and the Department of Energy's Role", Oak Ridge National Laboratory, March 2000.

[10] American Iron and Stee   l In stitute (AISI), S teel Recy cling Institu te," Recycling Scrapped Automobiles", http://www.recycle-steel.org/

[11] Helen N. Han, "The Environm   ental Impact of Steel and Alum   inum Body-in-Whites", *JOM*, February 1996.

[12] U.S. Environmental Protection Agency, "Em ission Facts: Average annual Emissions and Fuel Consum ption for Passe nger Cars and Light trucks", Office of Transportation and Air Quality, April 2000.

[13] The Aluminum Association, Inc., Automotive Aluminum, "Recyclability and Scrap value", http://www.aluminum.org/.

[14] Ada m Gesing, and Richard Wolanski,   "Recycling Light Metals from   End-of- Life Vehicles", *JOM*, November 2001.

Aluminum Alloys for Transportation, Packaging, Aeropsace, and Other Applications
Edited by Subodh K. Das, Weimin Yin
TMS (The Minerals, Metals & Materials Society), 2007

FABRICATION OF CARBON FIBERS REINFORCED ALUNIMU FOAM

Zhuokun Cao, Guangchun Yao, Yihan Liu

School of Material and Metallurgy of Northeastern University
Box 117, 110004, Shenyang, China

Keywords: Carbon fibers, Aluminum foam, Foam ability, Copper coating

Abstract

Carbon fibers reinforced aluminum foam was prepared by adding short copper coated carbon fibers and blowing agent TiH_2 in to aluminum melt. Influences of fiber content on foaming ability of the melt were discussed. Microstructure of the cell wall surface and cross section was observed by SEM to analyze the interface wetting ability and distribution of the fibers. Orientation of carbon fibers located in thin cell walls appeared to be parallel to the cell wall surface, which made reinforcement of the fibers more efficient. And suitable processing parameters, including stirring time, stirring speed and holding time were determined.

Introduction

Metallic foams offer attractive combinations of low density, high beam stiffness, good energy absorption and certain other properties [1]. Thus, they have attracted strong industrial and scientific interest during the last decade. Different methods have been developed to produce foams which can be divided into two categories: direct foaming by introducing gas bubbles into a conditioned melt and foaming with the help of blowing agents. Although the methods are apparently quite different, it is generally accepted that the foam ability of the matrix material is intimately correlated with the presence of particles, which is considered to increase the viscosity of the melt and/or to decrease the surface tension [2-4]. In practical manufacture, silicon carbide and alumna particles have been already used to increase the stability. It was also reported that the addition of these particles has effects on the mechanical properties of aluminum foam [1,5].

The addition of short carbon fibers in the melt could more efficiently increase the viscosity than particles, and improve the mechanical properties of the aluminum matrix [6]. However, little reports were found to take research on carbon fibers reinforced aluminum form, partially because that the fabrication of C_f/Al composites undergoes several difficulties: bad wetting ability at the interface and the formation of brittle phase and deterioration of the fiber properties during composite process. Many investigators have studied the carbon fiber/aluminum system, and it is reported that copper coating deposited on the fibers helps overcome these problems [7,8].

In this paper, copper coated fibers were used for improvement of the foam ability of molten aluminum, and the suitable processing parameters to fabricate carbon fibers reinforced aluminum foam is discussed.

Experimental Procedure

Preparation of Copper Coated Carbon Fibers

The carbon fibers used is 12K fibers offered by Anke company (Shenyang, China). Copper layer was electroplated on the fibers in an acid copper bath, details in [9], and the composition of the solution is listed in Table I.

Table I. Compositions of the Electroplating Solution

$CuSO_4$	120g/l
H_2SO_4	150g/l
Additive Agents	5ml/l

Fabrication of Aluminum Foam

Coated fibers were chopped to 3-5 mm at length, and added into molten aluminum at 700°C under a stirring speed of 800-2000rad/min. TiH_2 was used as blowing agent, and its content is 1.5wt%. After stirring and hold for a few minutes, the foam was taken out from the finance and cooled by air.

Characterization of the foams

The foaming ability of the melts with different content of carbon fibers is characterized by the expansion rate of the melt, which is obtained by measuring the volume of aluminum before and after foaming process. The content of carbon and copper in different part of the foam is determined by chemical analysis. And the selected parts for chemical analysis are shown in Figure 1.

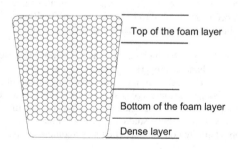

Figure 1. Selected parts of the foam for chemical analysis

Results and Discussion

<u>Wetting Ability of Coated Fibers</u>

SEM micrograph of fabricated composites before foaming process is shown in Figure 2. Uniform distribution of the fibers could be remarked. It is well known that if the wetting ability of the fibers towards molten aluminum is bad, the fibers would agglomerate under the surface tension of the melt. Therefore, it appears that the wetting ability at the interface is greatly improved by copper coating.

However, large quantity of CuO powders ware found inside the pores when adding in TiH$_2$ immediately after the addition of coated fibers. These CuO powders results from the oxidation of copper during foaming process, which indicates that most of the copper layer on the carbon fibers have not dissolve into molten aluminum. Thus, a certain period of holding time is needed before adding TiH$_2$ to insure that most of copper has dissolved into aluminum and avoid the oxidation of copper. The results of experimental study show that the suitable holding time after adding coated fibers is 10min.

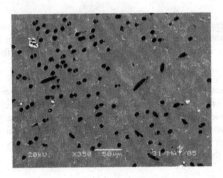

Figure 2. SEM micrograph of fabricated composites before foaming process

<u>Forming Ability</u>

Effect of fiber addition on the expansion of Al foams is shown in Figure 3. Since addition of fibers could increase the effective viscosity of the melt, the foam ability increased as the content of carbon fibers. However, there is a decrease of the foam ability when the content of fibers is over 3vol%. This is because that the viscosity of the melt becomes too high to adding in more fibers and the fibers can't be uniformly scattered in the molten aluminum.

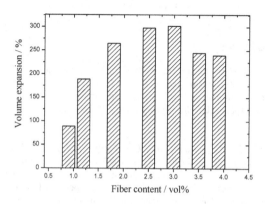

Figure 3. Effect of fiber addition on the expansion of Al foams

The content of carbon fibers and copper at different part of the foam is shown in Table II. It can be seen that the copper content is uniform at different layer, while carbon fibers mostly contain in the foam layer. And this indicates that it is the carbon fibers that contribute to the increase of foam ability.

Compared to aluminum foam of other kinds, the volume fraction of fibers is less than that of particles about 10μm size [10], which exhibits that short fibers are more effective on increasing effective viscosity of the melt. However, the most expansion rate of these foams is smaller than that of aluminum foam stabilized by particles [3,5,11]. This is because the bulk viscosity of the melt doesn't increase much and there are larger spaces between the fibers and thus the liquid is easier to drainage through the fibers under the effect of gravity.

Table II. Content of carbon fibers and copper at different part of the foam

Content of	Top of the foam layer	Bottom of the foam layer	Dense layer
Carbon	1.49	1.38	0.03
Copper	3.83	3.55	3.72

Cell Structure

As shown in Figure 4, the structure of the carbon fibers reinforced aluminum foam illustrates relatively uniform cells under a stirring rate of 800-1200rad/min, while lager irregular cells are seen at the cross-section of the foam when the stirring rate is much higher. SEM observation of the inner surface of these irregular cells, shown in Figure 5, reveals the fact that coating on some fibers was brush off after the foam process. Kaptay's discussion [12] and other studies have point out that the contact angle at the particle interface being able to stabilize liquid foam is at an uncertain range under 90°. However, it is evident that the contact angle at the interface between the naked carbon fibers and molten aluminum is over 120° at 700□. Therefore, when the copper layer on the fibers is brushed off, the foam can't be sufficiently stabilized by the fibers and turn to irregular. Thus, to gain a uniform cell structure, the stirring time is 800-1200rad/min.

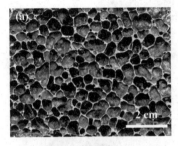 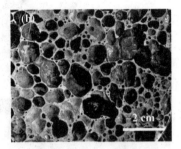

Figure 4. Optical macrographs show the cell structure of foam obtain under stirring of (a) 800-1200rad/min and (b) 1500rad/min

Figure 5. SEM micrograph of inner surface of the cell

Microstructure

The microstructure of plateau border and cell walls is shown in Figure 6. A uniform distribution of the fibers is observed, and the orientation of the fibers tends to be parallel to the nearby boundary of the cell walls. This orientation of fibers results from the liquid flow of the melt in the cell walls before solidification, which flows along the boundaries during the foaming process [13]. The orientation of fibers may also have effect on the mechanical properties of the foam, and further discussion would be taken in future work.

29

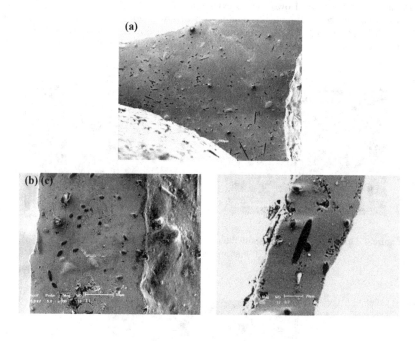

Figure 6. Microstructure of (a) plateau border and (b, c) cell walls

Conclusion

The fabrication of carbon fibers reinforced aluminum foam encountered several new challenges, including wetting ability at the interface and foam behavior of the new molten system. Through our discussion in this paper, we can conclude as follows:

1. Copper coating on carbon fibers could sufficiently improve the wetting ability at the interface and led to a uniform microstructure.

2. Carbon fibers perform the main effect on the foaming ability of the melt, and suitable volume fraction of the fibers is 2-3%.

3. The optimized processing parameters are as follows: holding time after adding in fibers, 10min; stirring rate, 800-1200rad/min.

4. The ordination of the fibers tends to be parallel to the nearby bonder of the cell walls, which results from the liquid flow of the melt in the cell walls before solidification.

Reference

1. John Banhart, "Manufacture, characterisation and application of cellular metals and metal foams," *Progress in Materials Science,* 46 (2001), 559–632.

2. Th. Wubben, S. Odenbach, "Stabilization of liquid metallic foams by solid particles," *Colloids and Surfaces A: Eng.Aspects,* 266(2005), 207-213

3. Carolin Korner, Michael Arnold, and Tobert F.Singer, "Metal foam stabilization by oxide network particles", *Materials Science and Engineering A,* 396(2005), 28-40.

4. S. Asavavisithchai, A.R. Kenndedy, "Effect of powder oxide content on the expansion and stability of PM-route Al foams," *Journal of Colloid and Interface Science,* 297(2006), 1331-1334.

5. S. Esmaeelzadeh, A. Simchi, and D. Lehmhus, "Effect of ceramic particle addition on the foaming behavior, cell structure and mechanical properties of P/M AlSi7 foam," *Materials Science and Engineering A,* 424(2006), 290-299.

6. T. W. Clyne, P. J. Withers, *An Introduction to Metal Matrix Composites,* (Cambridge: Cambridge University Press, 1993), 301-305.

7. J.F. Silvain, A. Proult, and M. Lahaye, "Microstructure and chemical analysis of C/Cu/Al interfacial zones," *Composites: Part A,* 34 (2003), 1143–1149

8. A. Urena, J. Rams, and M.D. Escaler, "Characterization of interfacial mechanical properties in carbon fiber/aluminium matrix composites by the nanoindentation technique". *Composites Science and Technology ,* 65 (2005), 2025–2038.

9. CAO zhuo-kun, LIU Yi-han, and Yao Guang-chun, "Electroplating of Carbon Fibers in Sulfate Acidic Solution," *The Chinese Journal of Process Enginerring,* 6(4)(2006), 651-655.
10. N. Babcsan, D. Leitlmeier, J. Banhart, "Metal foams—high temperature colloids Part I: Ex siu analysis of metal foams", *Colloids and Surfaces A: Physicochem. Eng. Aspects,* 261 (2005), 123-130.

11. V. Gergely, D. C. Curran, T. W. Clyne,"The FOAMCARP process: foaming of aluminum MMCs by the chalk-aluminum reaction in precursors," *Composites Science and Technology,* 63 (2003), 2301-2310

12. G. Kaptay, "Interfacial criteria for stabilization of liquid foams by solid particles", *Colloids and Surfaces A: Physicochem. Eng. Aspects,* 230 (2004), 67–80.

13. L. W. Schwartz, R. V. Roy, "A mathematical model for an expanding foam," *Journal of Colloid and Interface Science,* 264 (2003), 237-249.

Fabrication of Ultrahigh Strength Aluminum Alloy by the Route consisting of Solid Solution, Large Deformation and Ageing

Xiaojing Xu[1], Xiaonong Cheng[1]

[1]Institute of Advanced Forming Technology, Jiangsu University, 301 Xuefu Rd.; Zhenjiang 212013, P.R.China

Keywords: Aluminum alloy, Enhanced solid-solution treatment, Large deformation processing, Mechanical properties

Abstract

In present study, a fabrication route of the aluminum alloy with ultrahigh strength and moderate tensile ductility has been developed using 2024Al alloy as experimental materials. This fabrication route consisted of a low-temperature Large Deformation Processing (LDP) combined with a pre-LDP solid-solution treatment plus post-LDP a low-temperature ageing treatment. The present study also found that an enhanced solid-solution treatment prior to LDP would result in an obviously higher hardness and strength, and little deterioration to tensile ductility.

Introduction

Large deformation processing (LDP) has been proven to be very useful in improving strength of materials due to the introduction of structure defects [1, 2]. Enhanced solid-solution treatment also can improve mechanical properties of materials considerably due to the more full dissolution of coarse second phase [3]. It is of great interest to add these two strengthening effects to significantly improve the strength of materials, making them more useful.

The objectives of this study are (i) to establish a processing route to produce ultrahigh strength aluminum alloy and illustrate the effect degree of enhanced solid-solution treatment, LDP and post-LDP ageing treatment, and (ii) to examine the main mechanisms of the improvement of strength.

Experimental

Commercial 2024Al alloy with the nominal chemical composition (in wt %) of 4.26 Cu, 1.45 Mg, 0.71 Mn, 0.24 Si, 0.28 Fe, 0.067 Zn, 0.039 Ti, and balance Al was used as experimental materials. Prior to LDP, the alloy was solid-solution treated at 773 K for 1, 14 and 24 h respectively, using room-temperature water as quenching medium. After the solid-solution treatment, LDP was carried out immediately. The 1 and 14 h solid-solution treated 2024 Al was subjected to equal-channel angular pressing (ECAP [2]) deformation processing with an imposed equivalent normal strain of approximately ~ 0.5 at room temperature. The 24 h solid-solution treated 2024 Al was subjected to compression deformation with an imposed equivalent normal strain of approximately ~ 0.5 at liquid nitrogen temperature. The unLDPed 2024Al and the LDPed 2024Al were naturally aged at room temperature for at least three days; Among them, some was further subjected to low temperature artificial ageing treatment at 373 K for 48 h.

Microhardness was measured using Vickers microhardness tester with the load of 1.96 N and the loading time of 20s. Tensile properties was tested with the sample size of 3.0 mm × 1.5 mm cross-section and 5.0 mm gauge length, at an initial strain rate of 1×10^{-3} s^{-1} and at room temperature. X-ray diffraction (XRD) analysis was performed using a Rigaku diffractometer (model D/max-2500PC) with Cu Kα radiation ($\lambda = 0.154056$ *nm*).

Results and discussion

Hardness

Table I gives the Hv microhardness of the 2024Al alloys subjected to different processing. It can be seen that the enhanced solution treatment (i.e., 14 and 24 h solution treatment), the LDP and the post-LDP low temperature artificial ageing treatment all have the capability to improve hardness. LDP appears to be most effective. Their combination can increase hardness enormously, up to more than 200 Hv, an increase of about 45 % as can been seen by comparing No.7 specimen with No.2 specimen.

Table I. Microhardness of the 2024 Al alloys subjected to different processing

No.	Material states	Microhardness (Hv)
1	As-received	138.7
2	1 h solid solution- natural ageing	137
3	1 h solid solution- ECAP-natural ageing	177
4	14 h solid solution- natural ageing	147
5	14 h solid solution- ECAP-natural ageing	191.5
6	14 h solid solution- natural ageing-low temperature artificial ageing	157
7	14 h solid solution- ECAP- natural ageing-low temperature artificial ageing	201.5
8	24 h solid solution- natural ageing	149
9	24 h solid solution-compression-natural ageing	195.75

Tensile properties

Figure 1 presents the engineering stress-strain curves of the 2024 Al alloy subjected to different processing. It can be seen that the enhanced solution treatment (i.e., 14 h solution treatment) and the LDP both have the significant capability to improve strength. Low temperature artificial treatment can increases slightly tensile ductility while without any deterioration to strength. The combination of enhanced solution and LDP increases the yield strength of 2024 Al alloy up to 600 MPa, an considerable increase of about 43% as can been seen by comparing Figure 1a with Figure 1d. In addition to the strengthening, this combination allows the 2024Al alloy have a moderate level of tensile ductility (12.7 %). Besides these above mentioned, it is also worth noting that enhanced solution treatment resulted in an obviously improvement of strength while with little deterioration to tensile ductility (comparing Figure 1b with Figure 1d).

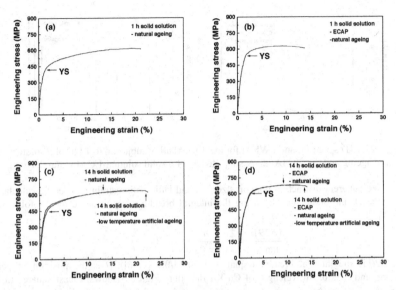

Figure 1. The stress-strain curves of (a) the 2024 Al alloys subjected to different processing

The contribution of dislocation increase to strength improvement

Figure 2 shows the XRD pattern and the full-width at half-maximum (FWHM) diffraction peaks for the 2024Al alloy subjected to 14 h solid solution-natural ageing and the 2024Al alloy subjected to 14 h solid solution-ECAP-natural ageing. With the comparison of Fig. 2a and Fig. 2c, it can be seen that the ECAP processing can decrease texture considerably, as can indicated by comparing their peak height ratio, which of latter is closer to the value for Al with randomly oriented grains. With the comparison of Fig. 2b and Fig. 2d, it can be easily found that the ECAP processing causes the XRD peaks to broaden considerably.

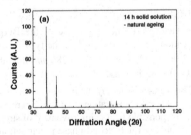
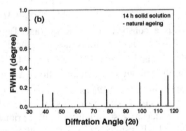

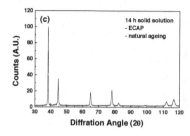 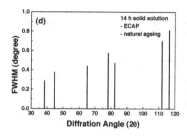

Figure 2. XRD spectrum and FWHM for the 2024Al alloy subjected to 14 h solid solution-natural ageing and the 2024Al alloy subjected to 14 h solid solution-ECAP-natural ageing

The average coherent diffraction domain size (d) and lattice microstrain (e) can be calculated from the XRD line broadening by using the integral breadth analysis based on the following equation [4]:

$$\frac{(\delta 2\theta)^2}{\tan^2\theta} = \frac{\lambda}{d}\left(\frac{\delta 2\theta}{\tan\theta\sin\theta}\right) + 25\langle e^2 \rangle$$

Where $\delta 2\theta$ is the integral breadth (equal to FWHM/0.9 [5, 6]), θ is the position of peak maximum, and λ is the wavelength of Cu Kα radiation. Figure 3 shows a least squares fit of the $(\delta 2\theta)^2/\tan^2\theta$ against $(\delta 2\theta)/(\tan\theta\sin\theta)$ for all measured peaks. The calculated average coherent diffraction domain size (d) and the root mean square lattice microstrain ($\langle e^2 \rangle^{1/2}$) were listed in Table II.

For the materials subjected to severe plastic deformation, dislocations density (ρ) can be represented in terms of d and $\langle e^2 \rangle^{1/2}$ by [7]:

$$\rho = 2\sqrt{3}\langle e^2 \rangle^{1/2}/(d \times b)$$

where b is the Burgers vector and equals to 0.286 nm for FCC Al. The calculated dislocation density is also listed in Table II.

The dislocation strengthening (σ_ρ) can be estimated by the Taylor relationship:

$$\sigma_\rho = M\alpha Gb\rho^{1/2}$$

where α is a constant, M is the average Taylor factor, b is the magnitude of the Burger's vector, G is the shear modulus, and ρ is the dislocation density. Substituting the typical values of α, M, G, and b for Al [8] (i.e., α=0.24, M=3.06, G=26 GPa, and b=0.286 nm) into above equation, the strengthening contribution ($\Delta\sigma_\rho$) due to the increase of dislocation density was calculated to be ~ 82.9 MPa as listed in Table II. This value divided by the improvement in yield strength (600-460) MPa produces a value of ~ 59.2 %, illustrating that the increase of dislocations density resulting from LDP is primarily responsible for the improvement of strength. In addition, LDP-resulted refinement of grain and relatively finer second-phase precipitation are considered to be another two main contributors of strengthening.

36

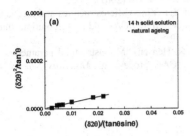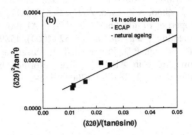

Figure 3. Integral breadth analysis to calculate the coherent domain size and lattice strain

Table II. Structural and mechanical features calculated from XRD data

	The 2024Al alloy subjected to 14 h solid solution-natural ageing	The 2024Al alloy subjected to 14 h solid solution-ECAP-natural ageing
Coherent domain size, d (nm)	65.0	29.1
Lattice microstrain, $\left\langle e^2 \right\rangle^{1/2}$ (%)	0.03174	0.1255
Dislocation density, ρ (10^{14} m^{-2})	0.591	5.23
Dislocation strengthening, σ_ρ (MPa)	42.0	124.9
Strengthening due to dislocation density increase, $\Delta\sigma_\rho$ (MPa)	0	82.9

Conclusions

A combination of pre-LDP enhanced solid-solution treatment, LDP at low temperature and post-LDP low-temperature aging treatment has the potential to render Al alloys significantly stronger than those processed by technique alone. The increase of dislocations density resulting from LDP is primarily responsible for the improvement in hardness and strength.

References

1. M.A. Meyers, A. Mishra, and D.J. Benson, "Mechanical properties of nanocrystalline materials", *Progress in Materials Science*, 51 (2006), 427-556.
2. R.Z. Valiev, R.K. Islamgaliev, and I.V. Alexandrov, "Bulk nanostructured materials from severe plastic deformation", Progress in Materials Science, 45 (2000), 102-189.
3. Kanghua Chen, Hongwei Liu, Zhuo Zhang, Song Li and Richard I. Todd. The improvement of constituent dissolution and mechanical properties of 7055 aluminum alloy by stepped heat treatments. *Journal of Materials Processing Technology*, 142 (2003), 190-196.
4. K.M. Youssef, R.O. Scattergood, K.L. Murty, and C.C. Koch, "Nanocrystalline Al–Mg alloy with ultrahigh strength and good ductility", *Scripta Materialia*, 54 (2006), 251-256.
5. T. Ungár, I. Dragomir-Cernatescu, D. Louër, and N. Audebrand, "Dislocations and crystallite size distribution in nanocrystalline CeO$_2$ obtained from an ammonium cerium(IV)-nitrate solution", *Journal of Physics and Chemistry of Solids*, 62 (2001), 1935-1941.
6. T. Ungár. Dislocation densities, arrangements and character from X-ray diffraction

experiments. *Materials Science and Engineering A*, 309-310 (2001), 14-22.

7. Y.H. Zhao, X.Z. Liao, Z. Jin, R.Z. Valiev, and Y.T. Zhu, "Microstructures and mechanical properties of ultrafine grained 7075 Al alloy processed by ECAP and their evolutions during annealing", *Acta Materialia*, 52 (2004), 4589-4599.
8. J.R. Bowen, P.B. Prangnell, D.J. Jensen, and N. Hansen, "Microstructural parameters and flow stress in Al–0.13% Mg deformed by ECAE processing", *Materials Science and Engineering A*, 387-389 (2004), 235-239.

Aluminum Alloys
for Transportation, Packaging, Aerospace and Other Applications

Aluminum Products

Aluminum Alloys for Transportation, Packaging, Aeropsace, and Other Applications
Edited by Subodh K. Das, Weimin Yin
TMS (The Minerals, Metals & Materials Society), 2007

IMPROVEMENT OF QUALITY OF TRIMMED SURFACE OF ALUMINUM PANELS

Sergey Golovashchenko

Ford Motor Company
MD3135, 2101 Village Road; Dearborn, MI 48124, USA

Keywords: Trimming, Aluminum Panels, Shearing Edges, Clearance.

Abstract

Trimming requires accurate alignment of the die shearing edges, typically 5-10% of the blank thickness. This clearance may be significantly larger due to insufficient stiffness of tooling and wear of the shearing edges. Increasing the clearance above the recommended value often leads to generation of burrs on the trimmed surface. These burrs may create difficulties for flanging and hemming operations. In addition, pieces of metal called slivers are generated during trimming process. Those pieces can be imprinted into the surface of stamped panels, which may require metalfinish of every stamped exterior panel. Also, imperfections of the trimmed surface may reduce formability of the blanks significantly below Forming Limit Diagram. A modified trimming process significantly reducing the described imperfections will be discussed. Conventional and modified process will be compared for various clearances and trimming angles. Aluminum alloy 6111-T4, often used for exterior automotive panels, was used in this experimental study.

Introduction

Modern product design and manufacturing often utilizes a wide variety of materials including aluminum alloys and advanced high strength steels. These materials often are capable of reducing vehicle weight; although, they may present difficulties when subjected to manufacturing processes originally designed for low carbon steel. One such manufacturing area where difficulties may arise is in trimming operations of automotive exterior body panels.

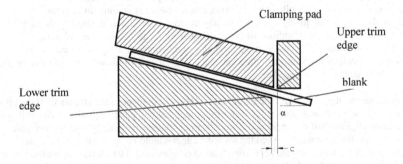

Figure 1: Schematics of the conventional trimming processes.

Similar to other shearing processes, the critical parameter for trimming is the clearance c between the shearing edges of the trim die, as it is indicated in Fig.1. Also, an important parameter is the cutting angle α, shown in Fig.1.

Die design literature [1] explains the mechanism of separation in shearing operations as a result of fracture initiation from both upper and lower cutting edges, as it is indicated in Fig.2 for blanking operation. If the clearance is suitable for the material being cut, these fractures will spread toward each other and eventually meet, causing complete separation.

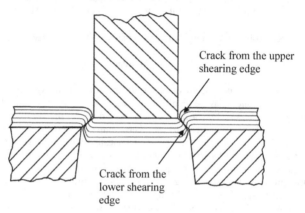

Crack from the upper shearing edge

Crack from the lower shearing edge

Figure 2: Mechanism of separation in shearing processes.

But the practical experience of stamping of aluminum body panels indicates that in many cases burs on the part side are generated and hair like pieces of aluminum are produced during the trimming process. These pieces, commonly referred to as slivers, are highly undesirable, since they may get attached to the blank surface and distributed to the dies following the trimming operation. The accumulation of slivers on both the die and blank surfaces can result in an unacceptable surface finish when the blank is subjected to press operations: the slivers located on either the die or the blank can be forced into the blank surface. This issue was a major problem during implementation of aluminum panels in high volume production. Another factor influencing the quality of panels directly in trimming operation is production of burrs. Traditionally, the overall quality of parts after any shearing operation is defined by the height of burrs on a sheared surface in addition to dimensional accuracy and absence of splitting, developing from trimmed surface in stretch flanging and hemming. Large burrs can create significant difficulties for hemming operation, since it is not easy to assemble two panels together, if one of them or both have large burrs.

Current standards attempt to limit the production of burrs through accurate alignment of the upper and lower edges for shearing operations like trimming, blanking, piercing, and etc. According to [1], accurate alignment of the upper and lower shearing edges is required to obtain acceptable surface quality: the clearance between the shearing edges should be less than 4.5-6% of the material thickness. Kozo Shibata [2] recommends a clearance set 0.04 - 0.08 mm irrespective to sheet thickness, while Atsushi Wanibuchi [3] sets a clearance between the shearing edges of 0 to 5% of the material thickness. In most cases industrially approved practice is a clearance of 10%

of the sheet thickness. However, in production practice the clearance may be significantly larger due to non-uniform distribution of the gap, wear of the shearing edges and flexibility of the trim steels. On production panels this can be evidenced by significant burrs, proportional to the clearance between the shearing edges [4]. M. Li, based on his experimental analysis [5], suggested to use inclined trimming, where cutting angle $\alpha \neq 0$ instead of perpendicular trimming. Suggested optimization of the process was done based upon the microstructural analysis and graphs of burr height as a function of the cutting angle, clearance between the shearing edges, and the radius of the moving shearing edge. Optimal trimming conditions, according to [5], can be achieved using cutting angle of about 15-20 degrees, dull shearing edges, and specific clearance range of 15-25% of the material thickness. However, in stamping practice usually all sides of the part require trimming operation and cutting angle is often dictated by the geometry of the part; therefore, arranging a specific cutting angle from all sides of the part is possible only if more expensive cam-trimming tools are employed. The objective of this paper is to suggest a new trimming process, which would eliminate sliver and burr generation and would lower the requirements to the accuracy of trim dies alignment.

Experimental technique

In order to simulate different trimming conditions, an experimental fixture was designed and constructed (Fig.3). The punch insert and die inserts were bolted to the upper and lower blocks correspondingly. Both blocks were mounted inside the die shoe providing the accurate alignment between the shearing edges. The clearance between them was adjusted to be uniform along the shearing line with the accuracy of about 0.01 mm. The clearance was varied by using a set of shims. Shims were positioned in between the upper block and the punch insert. Horizontal stiffness of the tool was increased by mounting an additional steel block on the lower plate. This block and the upper die block were adjusted to slide one along the other with almost no clearance when the press ram moves down.

Figure 3. Experimental tooling simulating variety of trimming conditions

When horizontal force is applied during the shearing process the steel block prevents the upper block from shifting to the right and increasing the clearance between the shearing edges. Aluminum sheet AA6111-T4 0.98 mm thick was used for this study. Sheets were cut into strips of 50 mm wide and 300 mm long. Each strip was sheared into samples about 12 mm long. Cross-sections of parts and scraps were ground, polished, and etched.

Mechanism of defects generation in conventional trimming process

To define the mechanism of sheared surface formation in conventional trimming process, we conducted a number of interrupted tests, which showed the initiation and development of original cracks from the upper trim steel. Even though both upper and lower steels have identical sharpness and produce approximately the same amount of strain during indentation into the blank, bending of the offal changes the overall symmetry of the shearing process creating additional tension near the upper shearing edge and additional compression near the lower shearing edge. It is known that almost every material has higher ductility in the compressive stress state than in the tensile stress state. This provides a qualitative explanation of the preferential development of cracks from the stretched area near the upper shearing edge as compared to the compressed area near the lower shearing edge. This statement is confirmed by cross-sections of part side of the sheared surface for variety of gaps exceeding 10% clearance between the shearing edges [4]. Bending of the offal takes place throughout the range of gap variation, even for the smallest values. More detailed study of the fracture mechanism indicated that the initial crack starts a small distance away from the sharp corner of the upper trim steel. The area around this sharp corner is subjected to large plastic deformation. The deformation significantly exceeds the total elongation, typically found in a tensile test for this material. This is possible due to compression of the material around the shearing edge. When the offal has been bent to a certain angle, some area of the blank goes out of contact with the die shearing edge. The small increment of deformation without any compressive pressure spends all the material ductility, remaining after the initial indentation of the shearing edge into the blank body. This process results in initiation of the crack from this zone and generation of the "tongue" on the top of the offal side of the sheared surface. During fracture development, the offal is bent down and the "tongue' is subjected to the horizontal forces from the vertical wall of the upper trim steel (Fig.4). These forces break the tongue off the scrap and generate the hairlike sliver, shown in Fig.5 on the front view of the sheared surface of the offal.

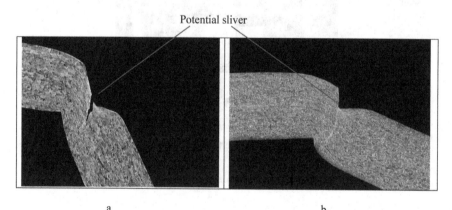

Potential sliver

a   b

Figure 4. Partial trimming with a) $\alpha=15°$ and b) $\alpha= 0°$

A detailed study of burr and sliver generation in trimming indicated that sliver may be generated in wide variety of clearances due to the nature of conventional trimming process, as it was indicated above. In Fig.6, the cross-sections of trimmed samples (α=15°) from both part and offal sides are illustrated. It confirms that with the increase of the clearance, the burr on the part side is growing, which requires rather accurate alignment of the shearing edges of the trimming die.

sliver

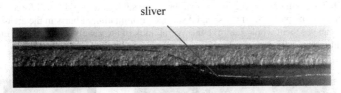

Figure 5. An example of a sliver on the offal side of the trimmed blank.

Cross-sections of slivers can be seen in Fig.6 for various gaps (20%, 30%, 40% and 50% of the material thickness) similarly as it was shown in Fig.4. Comparing the cross-sections of the offal for different clearances, it can be indicated that the angle of bending of the offal β before it separates from the scrap is growing with the increase of the clearance between the shearing edges. More detailed explanation is provided in Fig.7, where the angle β is shown as an angle between the area of the blank under the upper shearing edge (shown dashed in Fig.7) and the upper boundary of the blank (shown as a solid line) outside the zone of plastic deformation, which undergoes rotation as a rigid body. As it can be seen from Fig.7b, the angle of bending β can be read from the offal profile.

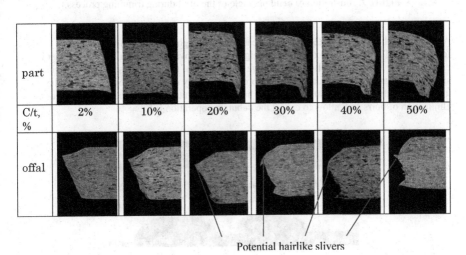

Potential hairlike slivers

Figure 6. Cross-sections of part and offal sides of trimmed surface produced in conventional trimming with specific clearances and cutting angle α=15°.

45

The burr on the part side due to complicated geometry of real panels and guillotine mechanism of trimming in order to reduce the required force of the press, may create uneven burrs. An example of such a burr is shown as a front view of the trimmed surface in Fig.8. Uneven burrs create two different problems: they might lead to splitting and, therefore, reduced formability of panels in stretch flanging and hemming; they also may get separated from the part creating a sliver of different origin, than the hairlike sliver shown in Fig.5. Burrs can also get separated from the trimmed part if there was a significant gradient of clearance along the trimming line. Due to the guillotine mechanism of shearing, these local burrs are subjected to additional forces, and they can be torn off from the part side of the sheared surface. Elimination of burrs in the areas, where customers can see them would improve the overall look of the panels. It may make unnecessary to use additional plastic parts, covering non-hemmed edges of the panel.

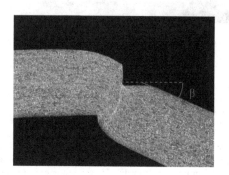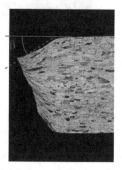

Figure 7. Illustration of bending angle of the offal during trimming process.

According to the graph in Fig.9 [1], the increase of burr height can dramatically reduce formability. Therefore, reduction of burr height may improve the quality of parts, where it is exposed to the customer. It also may improve formability of trimmed parts during following stamping operations. In addition, it may improve the quality of hemmed joints between the interior and exterior panels. In cases, where burrs are absolutely unacceptable, reduction or even elimination of burrs may allow to cancel the deburring operation [6].

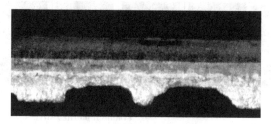

Figure 8. An example of non-even burrs on the part side of the trimmed blank

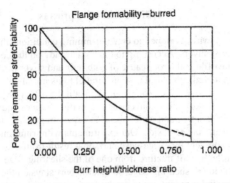

Figure 9. The effect of burr height on material formability.

The effect of the cutting angle α was studied in a separate set of experiments, using the same basic tool, shown in Fig.3. It required a special set of lower trim tools with appropriately machined lower shearing edges. The cutting angle was varied at four levels: 0°, 15°, 30° and 45°. This range of the angle α variation covers majority of practical cases of trim dies. The cross-section of part and offal sides after trimming with 30% clearance is shown in Fig. 10. It can be indicated that the quality of trimming is getting worse with the growth of the cutting angle. Further increasing of the cutting angle beyond 45° is problematic, unless cam mechanism for the upper trim tools is employed.

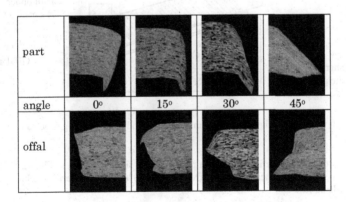

Figure 10. Cross-sections of part and offal sides of trimmed surface produced in conventional trimming with the clearance of 30% and various cutting angles α.

Based upon these considerations, the direction of modification of trimming technology should be towards prevention of scrap bending and separation of hair-like slivers due to rotation of the offal. This modification can be accomplished by adding the offal mechanical support to the existing concept of trimming.

47

Modified trimming process

A new trimming process was developed to create trimming conditions that eliminate burrs and prevent the generation of slivers. According to the above study on the conventional trimming process, bending of the offal is the root cause of both burr and sliver formation. In order to prevent these defects, we need to make the offal motion parallel to its initial position without rotating. This can be achieved by putting an elastic pad underneath the offal [7,8], as shown in Fig.11. Putting an elastic offal support into a conventional trim die provides equivalent conditions for fracture initiation from the upper and lower shearing edges. In practice this may result in pieces of burrs on both part side and the offal side. Our experiments with trimming strips confirmed this conclusion. In order to prevent the piece-wise burrs that result from dual mechanism of fracture, we need to create a preference of fracture from one of the shearing edges. It is preferable to have the crack start from the lower shearing edge, so that the burr stays on the offal. In this case the dimension of the part is independent of the gap between the shearing edges, which is the reason for expensive tool alignment in the conventional trimming process. To achieve this effect, we should decrease the concentration of strains on the moving shearing edge by dulling the upper trimming steel (Fig.11). This will lower the maximum level of strain at this shearing edge and create the preference for the fracture development from the lower trim steel. The radius of the shearing edge of the moving steel is a matter of adjustment. In our experimental study, the radius of 0.12 mm produced excellent results in trimming a sheet of Al 6111-T4 0.93 mm thick. For such a process, the range of clearances where trimming is possible is much wider than for standard trimming schemes.

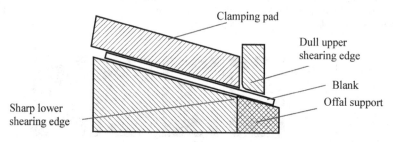

Figure 11: Schematic of the modified trimming process

An example of sheared surface from both part and offal sides is shown in Fig. 12 for the range of clearances between the shearing edges from 2% to 50% of the material thickness and for the cutting angle of 15°.

Comparison of samples in Fig.6 and Fig.12, indicates that modification of the trimming conditions may significantly improve the quality of trimmed surface and further performance of the trimmed parts. Since quality of the trimmed surface is mostly equivalent through all the range of gaps 2%-50%, the clearance between the shearing edges may be varied in much wider range. Such flexibility of the process makes upper and lower die alignment significantly easier. Therefore, we call this process robust trimming. Parallel movement of the blank, clamped between the upper trim tool and the offal elastic support, in modified trimming process eliminates bending of the offal and also eliminates the horizontal forces, which cause separation of the "tongue" from the offal side in the conventional trimming process. Therefore, the modified process has a strong potential to eliminate sliver defect, and, as a result, significantly reduce metalfinish work and produce cleaner class A surface. Improved quality of trimmed surface in modified trimming process was also confirmed by a set of experiments (Fig.13) with 30% clearance between the shearing edges and the cutting angle α varied at four levels: 0°, 15°, 30° and 45°, similarly to how it was done for the conventional trimming process. It can be indicated that significant improvement of quality of trimmed surface may be obtained for rather wide range of trimming angles variation, covering majority of practical applications.

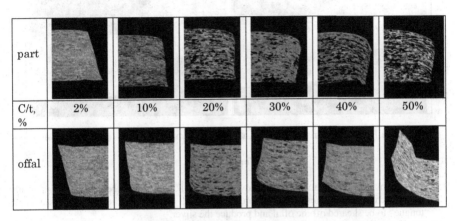

Figure 12. Cross-sections of part and offal sides of trimmed surface produced in modified trimming with elastic offal support with the cutting angle α=15°.

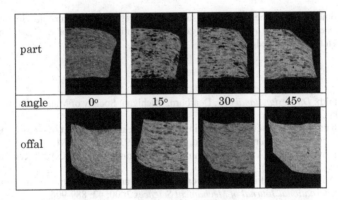

Figure 13. Cross-sections of part and offal sides of trimmed surface produced in modified trimming with elastic offal support with the clearance of 30% of material thickness and various cutting angles α°.

As it was mentioned above, the modified trimming process was engineered to have crack propagation to start from the sharp lower shearing edge. In order to confirm this mechanism of separation, several tests were done with partial shearing, where the separation process was not completed. Two of partially sheared samples are shown in Fig.14, where crack propagation from the lower trim steel is very well visible. The initial clearance between the shearing edges was.

Crack propagation

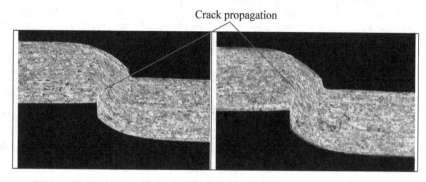

Figure 14. Mechanism of separation at modified trimming process ($\alpha= 0\,^\circ$, c=50%)

Conclusions

1. Experimental study of trimmed parts showed that hairlike slivers can be generated as a result of the fracture development from the upper trim steel. Bending of the offal causes a small "tongue" to be sheared off the offal and produce the sliver.

2. Mechanical support of the offal prevents its bending and eliminates separation of potential slivers by horizontal forces.

3. The combination of the mechanical support and preferential fracture development from the lower shearing edge provides robustness of the trimming process for wide range of clearances between the shearing edges.

References

1. D.A Smith: *Die design handbook*, Dearborn, Michigan: Society of Manufacturing Engineers. (1990)
2. K. Shibata: Clearance setting method in punching shear work, Patent 61-46329, Japan, 1986.
3. A. Wanibuchi: Work punching method, Patent 59-66934, Japan, 1984.
4. S.F. Golovashchenko: A study on trimming of aluminum alloy parts, Advanced Technology of Plasticity, Vol. 3(1999), p.2261-2266.
5. Li M. An experimental investigation on cut surface and burr in trimming aluminum autobody sheet. *International Journal of Mechanical Sciences* 2000; 42: 889-906.
6. Gillespie LK. Deburring and edgefinishing handbook. Dearborn, MI: Society of Manufacturing Engineers, 1999.
7. S.F. Golovashchenko, L.B. Chappuis, G. Baumann, R. Regelin, G. Masak., R.R. Weir and W.S. Stewart: Apparatus for trimming metal. US Patent Publication US 2003 0209536A1, 2003.
8. S.F. Golovashchenko, M. Blodgett: Apparatus for Trimming Metal, Patent Application # US2005/0022640A1, 2004.

Failure Analysis of Some Aluminium Alloys Used to Produce Aluminium Containers Especially Three Compartment and Airline Trays

Authors: Nick Hall Singleton, Mohamed Essa, Ebrahim Al-Aradi, Nicasio L. Velante

Affiliations: Gulf Aluminium Rolling Mill Company; P.O. Box 20725; Kingdom of Bahrain

Keywords: Semi Rigid Container (SRC), Thermal Softening Curve (TSC), Formability, Anisotropy

Abstract

Aluminium is a widely used metal in various branches of industry like packing, construction, aerospace, etc. In this work, cracks during the formation of some semi-rigid foil food containers, especially the three compartments and airline casserole trays are investigated in aluminium alloys from the 3XXX and 8XXX series with different iron contents. It has been found that higher iron content up to 1.6 % shows significant improvement in yield strength and improved formability (even if the die design is poor). This paper will focus on the influence of the iron/silicon ratio and its strengthening effect on formability to prevent cracking during the pan forming process.

Introduction

Aluminium is a widely used metal in various branches of industry: packing, construction, aerospace and automotive industry to mention just a few areas. Aluminium properties can be improved by adding other elements to maximize workability or corrosion resistance. Various elements such as magnesium, iron, silicon and manganese and copper are added to the molten aluminium. During the solidification process, segregation of these alloying elements takes place and small precipitate particles, which are rich in alloying elements, are formed. Casting structure obtained will be un-homogenized or heterogeneous structure and this will affect the properties when the metal is subjected to other processes. These particles will give unwanted local variations of the mechanical properties and this limits the applicability and the efficiency of the mechanical treatments. To remove these compositional variations, and hence improve its mechanical properties, the alloy is heated up to temperatures just below the eutectic temperatures and kept at this temperature for some hours. This process is called homogenization and is usually applied to wrought aluminium non heat treatable alloys such as 1xxx, 3xxx, 5xxx and 8xxx. These alloys are strengthened by alloying elements in solid solution and dislocation structures introduced by cold rolling. GARMCO FOIL uses 1xxx, 3xxx, 5xxx and 8xxx series for products like household foil, food containers, fuel cans, platter tray, blade cutters, adhesive tape, insulation, lids, lamination and fin stock.

This paper focuses on the influence of the iron silicon phase on the mechanical properties, the strengthening mechanism of iron and how the formability is improved to prevent cracking during forming of three compartment and airline trays. The main objectives of this work are to study the effect of iron on the formability of aluminium foil, to study the effect of iron on the microstructure and to evaluate the changes in the yield strength as the iron content increases.

Recently, the 8xxx series alloys have become widely used in the Middle East market. Some alloy grades contain as much as 2.0% Fe with small amounts of other elements. They have excellent properties in strength, elongation and ductility. The 8xxx series has been mainly developed for packaging and is successful due to good formability. The important properties of this series can help solve the cracking problem with some food containers, especially the three compartment and airline casserole trays. When other alloys are used for the same application, the rate of failure is higher. Another point is the failure cases that occur are not because the other alloys are weak; in fact they can have more strength. It is due to the second phase which creates local strain leading to cracks around that spot. In the case of 8xxx series in H2X tempers, the Fe/Si phase gives more reliability in all directions and the aluminium will withstand high stresses introduced by punching during forming of semi rigid containers (SRC). The formability of each alloy is directly dependant on its chemical composition and the formation of intermetallic compounds during casting. These compounds play an important role of strengthening the alloy during rolling and heat treatment.

This paper aims to study the effect of chemical composition, especially the iron content, and how it changes the properties of the aluminium, improves the formability and reduces cracking even when using the same die for the down gauged material.

Discussion & Results

The aluminium industry in Bahrain started in the 70's with the opening of Aluminium Bahrain (ALBA). ALBA is the first smelter opened in the Gulf region with a starting capacity of 300 KMT/year; the capacity is now increased to reach over 800 KMT/year of ingots, billets and rolling slabs. After ALBA started production, downstream aluminium industries were established. The major downstream industries are GARMCO, MIDAL CABLES and BALEXCO. Our rolling division was established in 1981 and we are the biggest downstream industry on the island. Our foil division was established in 1999.

Metallurgical Aspects of Aluminium

Aluminium and its alloys are the most formable of the common fabricated metals. There are distinct differences in chemistry, homogenization, hot rolling, annealing and cold rolling processes between aluminium alloys that affect formability. Formability is the extent to which the metal can be deformed under a particular process before failure occurs. This failure is usually due to either localized necking or ductile fracture. The most common example where cracking failures occur is at severe drawing and stretching near the bottom corners of SRC. Cracks are the major problem resulting in customer complaints. See Figure 1 below.

Figure 1: Three compartment tray with severe cracking.

Alloys used for aluminium SRC are 1100, 3003, 3004, 3005, 8011 and 8006. Each alloy has a definite chemical composition as per Aluminium Association Standard [1].

Background to the Forming Issues

Alloys 1100, 3003, 8011, 8079 and 8006 are successfully being used in the Middle East. Alloys 1100, 8011, 8079 and 8006 are being used for household foil applications at 11 to 18 microns. No forming issues occur in this application. As the price of aluminium increased, cost reduction in the form of gauge reduction took place in household, SRC and fin stock foil products. Light gauge household foil was reduced from 14 to 11 microns, a 21% gauge reduction. Three compartment trays were reduced from 95 to 68 microns, a 28% gauge reduction. Fin stock was reduced from 150 to 115 microns, a 23% gauge reduction.

This reduction occurred over a 3-year time span and was not a serious forming or cracking issue until the gauge reductions reached 10%. At this point in the cycle, the cracking failures became a serious issue. Once the cracking failures increased over 3% SRC failures, the cost reduction economics did not apply. The scrap rate was costing more than the savings in gauge reduction. Our customers began looking for better formability and reduced scrap rates. They found alloy 8006 supplied out of Italy and Turkey to give excellent performance. Now the pressure is on us to be competitive.

Alloy 8006 was developed as a continuous cast (CC) product and has been converted to direct cast (DC) by most producers. Many DC operations moved away from 8006 alloy due to safety issues of bleed outs causing DC pit molten metal explosions and quality issues of large intermetallic particles. Another quality issue has been increased die wear in the fin stock industry that required special dies for 8006 alloy.

Samples and information collected from our customers produced the following information: Alloy 8006 in H22 temper is being used for critical forming application in both fin stock and SRC applications. Alloy 1100 (O to H2X tempers) has serious failures in fin stock and 3003 (O to H2X tempers) has 3 – 10% cracking in SRC.

With 3003 alloy, cracking almost always occurs in the transverse rolling direction. SRC Alloy 8006 in H24 temper has excellent elongation properties 90 degrees to the rolling direction. Table II below shows the competitor's 8006 alloy compared to GARMCO's 3003 alloy in H24 temper.

Table II. Mechanical Properties for Competitor's 8006 Alloy vs. GARMCO's 3003 Alloy in H24 Temper.							
Transverse		Direction			Rolling Direction		
Alloy Gauge mm		UTS Kg/mm²	YS Kg/mm²	% Elong.	UTS Kg/mm²	YS Kg/mm²	% Elong.
8006	0.052	14 12 13			14 11 14		
3003	0.052	14 11 9			15 11 14		
8006	0.055	14 10 16			14 10 15		
3003	0.055	15 11 9			13 11 15		
8006	0.075	14 10 23			14 11 19		
3003	0.075	14 10 11			14 10 17.0		

Alloys 8011 and 8006 are being used for household foil applications.
The mechanical properties of 8006 in O temper do not appear to have an advantage over 3003 in O temper. However, during the forming operation, 8006 has less cracking.
After collecting the above information, we decided to produce trials that would successfully determine how to produce SRC products without cracking failures.

3003 Alloy Trials

The first application of 3003 alloy to three compartment trays was successful at 95 micron gauge. As the gauge was reduced to 85 microns, cracking occurred. The cracking occurred in the areas subjected to high stresses and was related to the alloy, mechanical properties, die design and lubrication. Die design is critical to successful forming. This was demonstrated to us when our customer reworked the 0.080 x 290mm die (the original gauge was 0.095 mm for this die). Before the rework, the 3003 O temper failed to make the pans. After the rework, the same alloy and temper was successful. The trial shown in Table III below at 80 microns was successful due to the customer reworking the dies that were originally designed for thicker gauges, to work at thinner gauges. Gauge reduction continues and is now at 68 microns. From customer feedback, 3003 alloy failed when applied to certain types of SRC where the die design caused localized strain areas that led to severe cracks. Table III below documents these trials.

Table III. Mechanical properties of the 3003 alloy trials.					
Alloy Temper	Size (mm)	UTS MPa	YS MPa	Elong. %	Feedback
3003 O	0.080 x 330	120	46	16.0	Failed
3003 O	0.080 x 290	124	51	19.3	Pass
3003 O	0.075 x 290	130	62	15.1	Failed
3003 O	0.075 x 290	120	45	18.0	Failed
3003 O	0.072 x 335	130	74	13.4	Failed
3003 H22	0.073 x 325	127	53	20.0	90% failed
3003 H22	0.075 x 290	119	60	12.9	Failed

8011 Alloy Trials

Samples were prepared from 8011 alloy and sent to different customers for their evaluation. The chemistry and mechanical property details of these trials are shown in Table IV and Table V below. This alloy trial was another failure with mixed results due to cracking at the corners and at the curl around the top edges. Our customers did not like the soft feel of pans from this alloy. 8011 alloy did not have the rigidity of 3003 alloy.

Table IV. Mechanical properties of the 8011 alloy trials.					
Alloy Temper	Specification (mm)	UTS MPa	YS MPa	Elongation %	Result
8011	O	0.080 x 340	90	35	21.7	Fail
8011	O	0.090 x 330	90	33	21.0	Fail
8011	O	0.080 x 360	92	35	23.2	Fail
8011	O	0.080 x 343	90	35	21.7	Pass
8011	O	0.075 x 300	87	32	22.2	Pass
8011	O	0.080 x 290	91	43	17.8	Pass

Table V. Chemical compositions of the samples produced for 8011 alloy.						
Alloy	Al	Si	Fe	Cu	Mn	Others
8011	98.34	0.746	0.714	0.057	0.056	0.038
8011	98.46	0.760	0.680	0.070	0.000	0.009
8011	98.47	0.780	0.660	0.060	0.000	0.009
8011	98.50	0.730	0.660	0.070	0.020	0.009

1100 & 8079 Alloy Trials

Trials were produced from 1100 and 8079 alloys. Table VI shows the mechanical properties of these alloys and 8011 when receiving the same amount of cold work. Alloy 1100 has higher UTS and YS than 8079 and 8011 alloys.

Table VI. Mechanical properties (as-rolled) vs. Fe/Si ratio				
Alloy type	Fe/Si ratio	UTS (MPa)	YS (MPa)	% EL
1100 4.23		184	166	3.0
8011 0.92		167	147	3.2
8079 5.60		176	157	2.0

Trials were produced from these alloys with various mechanical properties in O temper and were sent to our customers for evaluation. Table VII shows the mechanical properties of the metal sent to our customers.

Table VII. Mechanical properties for the O temper trials.				
Alloy type	Fe/Si ratio	UTS (MPa)	YS (MPa)	% EL
1100 4.23		88	30	25.0
8011 0.92		89	31	31.0
8079 5.60		95	32	35.0

The results varied from customer to customer using the same die design. Full success was not achieved due to cracking. However, we noted that the 8079 alloy had fewer cracking failures than alloys 8011 and 1100 alloys. Alloy 1100 had more failures than 8011 alloy. Our customers were not impressed with SRC rigidity from these alloys.

8006 Alloy Trials

From the evaluations of all trials, we determined that another alloy system would be selected for evaluation. 8006 alloy was selected to obtain better formability and good rigidity. One cast was made using 4 Wagstaff 460 x 1650 mm moulds for evaluation. The chemistry is shown in Table VIII below.

Table VIII. Chemistry of the 8006 alloy trial.											
Al	Si		Fe	Cu	Mn	Mg	Cr	Ni	Zn	V	Ti
97.60	0.260		1.64	0.053	0.38	0.005	0.003	0.003	0.0003	0.114	0.003

This alloy has good formability properties. See the trial results in Table IX below.

Table IX. Feed back on experimental samples prepared from 8006 alloy.					
			Mechanical Properties		
Alloy/Temper	End Use	Feedback	UTS (MPa)	YS (MPa)	% EL
8006 O	Multi cavity	Success	116	76	21.0
8006 O	Airline tray	Success	113	73	26.0
8006 O	Fin stock	Success	112	83	30.0
8006 O	SRC	Success	114	63	24.0

With the success of 8006 alloy in O temper, we produced more trials in the H2X tempers for SRC and household foil and these also successful. See Figure 3 below for a typical thermal softening curve (TSC) for 8006 alloy. Note the good anisotropic mechanical properties in various tempers from as-rolled to full anneal. The tensile strength (TS), yield strength (YS) and elongation are almost the same for the rolling direction as compared to the transverse direction.

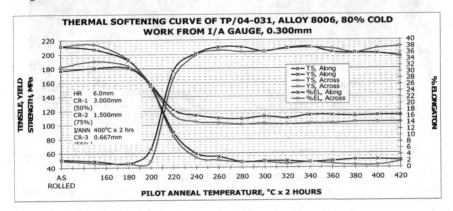

Figure 3: Typical TSC for 8006 alloy.

This information was used to develop the annealing practices for the H2X tempers for each trial gauge.

8079 Alloy Higher Fe Trials

Table X below shows the effect of iron content on the properties of 8xxx alloys at 0.075 mm finish gauge.

Table X. Effect of Iron Content on the Properties of 8xxx Alloys						
Alloy	Fe %	State	UTS, MPa	YS, MPa	El %	Erichsen (mm)
8011 0.65		As Rolled	163	143	3.7	1.85
		Annealed 80		28	36	7.75
8079 1.07		As Rolled	177	155	3.2	1.80
		Annealed 86		34	36	6.35
8006 1.64		As Rolled	216	177	2.1	1.75
		Annealed 110		61	34	5.75

With the success of 8006 alloy, we decided to revisit 8079 alloy and increase the Fe content from 1.0% to 1.2%. The successful results are shown in Figure 5 below. The cracked SRC on the left has 1.0% Fe. The good SRC tray on the right has 1.2% Fe.

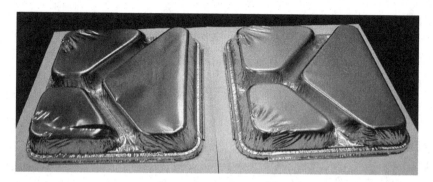

Figure 5: Three compartment trays prepared from 8079 alloy with different Fe contents.

The mechanical properties for the 8079 alloy trials are shown in Table XI below. Note the Erichsen increase from 4.3 to 5.8 mm and the elongation increase from 18 to 23% for 8079 with 1.2 % Fe.

Table XI. Mechanical Properties of Experimental Samples Produced from 8079 Alloy in H22 Temper.						
Alloy	Fe %	Gauge (mm)	Erichsen (mm)	UTS (MPA)	Y.S (MPa)	% EL
8079 1.04		0.066	4.3	10	7	18
8079 1.20		0.068	5.8	11	7	23

Table XII below summarizes the results for all alloys by customer evaluations. These results show that 8006 alloy has the best customer evaluations. Note that the successful 3003 O temper trial at 0.080 x 290 mm was with refurbished dies.

Table XII. Summary of Mechanical Test Data for All Alloys							
Alloy Temper		Fe/Si Ratio	Size, mm or End Use	UTS (MPa)	YS (MPa)	% EL	Remarks
3003 O			Multi cavity	120	46	16.0	Fail
3003	O		0.080 x 290	124	51	19.3	Pass
3003	H22		0.073 x 325	127	53	20.0	90% fail
3003	H22		0.075 x 290	119	60	12.9	Fail
8011 O		0.92	Multi cavity	90	35	21.7	Fail
8011 O		0.92	Multi cavity	90	35	21.7	Pass
1100 O		4.23	Multi cavity	88	30	25.0	Most failures
8079 O		5.60	Multi cavity	95	32	35.0	Few failures
8006 O		6.33	Multi cavity	116	76	21.0	Pass
8006 O		6.33	Airline tray	113	73	26.0	Pass
8006 O		6.33	Fin stock	112	83	30.0	Pass
8006 O		6.33	SRC	114	63	24.0	Pass

The recrystallized grain structures for 1100, 3003, 8077, 8079 and 8006 alloys at the same finished gauge and percent cold work are shown in Figures 6 through 10 below. These microphotographs were supplied by SECAT [2]. The average American Society for Testing Materials (ASTM) grain size is shown in Table XIII below [3]. In our process, 8006 alloy has the smallest grain size.

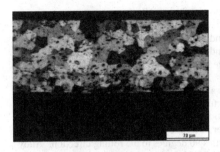

Figure 6: Optical photomicrograph of 1100.

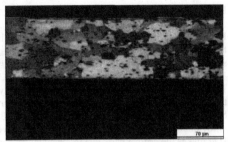

Figure 7: Optical photomicrograph of 3003.

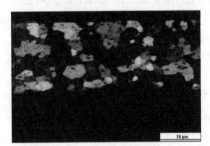

Figure 8: Optical photomicrograph of 8011.

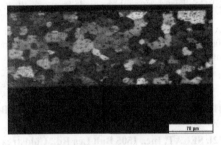

Figure 9: Optical photomicrograph of 8079.

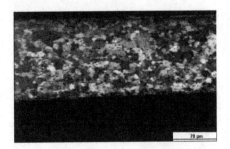

Figure 10: Optical photomicrograph of 8006.

Table XIII. Average ASTM Grain Size	
Sample ID	Average Grain Size
1100 8.52	
3003 8.73	
8011 8.89	
8079 9.08	
8006 11.06	

Conclusions

From the trial results, the following conclusions can be made:
Successful SRC forming requires three important combinations, the right selection of alloy, temper, die design and surface lubrication.

Down gauging without die modification is a recipe for cracking SRC.

From the trials done, it is clear that each alloy has individual physical and mechanical properties that are due to the influence of the phases present and their concentrations. The principal strengthening agents in the 8xxx series alloys are the Fe – Si constituent particles.

The material properties during forming are strongly affected by the rate of strain hardening and the strain rate. As the iron content increases, better formability occurs and successful forming of SRC is realized even under the most difficult conditions of die complexity or down gauging. Alloy 8006 has improved anisotropic properties over 3003 alloy in H2X tempers. Alloy 8006 H22 makes the critical parts due to excellent properties 90 degrees to the rolling direction. Alloy 3003 usually always fails in this direction. Alloy 8006 H22 does make the SRC in the applications were 3003 fails.

Alloy 8006 has the potential advantage in a foil rolling operation since fin stock, household and SRC foil products could be made from one alloy.

Alloy 8006 successfully makes all fin stock applications where 1100 alloy does not. Alloy 1100 fails in 10 – 15% of the applications were the part has been seriously down gauged without die modifications.

References

[1]. The Aluminum Association, 1525 Wilson Boulevard, Suite 600, Arlington, VA 22209, www.aluminum.org.

[2]. SECAT, Inc., 1505 Bull Lea Rd., Coldstream Research Campus, Lexington, KY 40511 ("SECAT"), www.secat.net. This is unpublished and confidential report for GARMCO.

Revision 1: 25 June 2006

[3]. American Society for Testing Materials, 100 Barr Harbor Drive, PO Box C700, West Conshohocken, PA 19428-2959, www.astm.org.

Aluminum Alloys for Transportation, Packaging, Aerospace, and Other Applications
Edited by Subodh K. Das, Weimin Yin
TMS (The Minerals, Metals & Materials Society), 2007

ALUMINUM ALLOYS FOR BRIDGES AND BRIDGE DECKS

Subodh K. Das and J. Gilbert Kaufman,
Secat, Inc., 1505 Bull Lea Road, Lexington, KY 40511

Keywords: Aluminum Alloys, Bridges, Design, Construction

Abstract

Aluminum alloys have been used in bridge structures since 1933, when the first aluminum bridge deck was used to replace an earlier steel and wood deck on Pittsburgh's Smithfield Street Bridge in order to increase its live-load carrying capacity. While still not considered a standard for bridge structures, aluminum alloys have much to offer for such applications, and continue to be used where their light weight, high strength-to-weight ratio, and excellent corrosion resistance satisfy service requirements.

This paper will provide in detail the advantages and limitations of aluminum alloys for bridge structures, including the key factor that they never require painting or any other type of coating for corrosion protection. A brief overview of the history of the use of aluminum in bridges in the United States and other locations throughout the world will also be presented.

Introduction

Aluminum alloys have been used in bridge structures for more than 70 years. In 1933, the first aluminum bridge deck was used to replace an earlier steel and wood deck on Pittsburgh's Smithfield Street Bridge in order to increase its live-load carrying capacity. Since that time aluminum has been used in various ways in hundreds of bridge structures around the world, and most remain in service today, including some for more that 50 years.

While still not considered a standard for bridge structures, aluminum alloys have much to offer for such applications, and continue to be used where their light weight, high strength-to-weight ratio, and excellent corrosion resistance satisfy service requirements and justify the additional initial cost. When considered on a life-cycle cost basis, aluminum bridges components have clear superiority.

The advantages of aluminum for bridges

Aluminum alloys have several important performance characteristics that make them very attractive for bridge structures, namely:

- Light unit weight, only one third that of steel;
- Strengths comparable to typical bridge steels;
- Excellent corrosion resistance, with negligible corrosion even in the presence of rain and road salts;
- High toughness and resistance to low-ductility fracture, even at very low temperatures, and free of any ductile-to-brittle transition that has sometimes been fatal to older steel bridges; and
- Excellent fabricability, including ease of production of extrusions to complex hollow shapes optimized for structural design and assembly

These performance characteristics provide significant advantages over conventional steel and concrete in the design, fabrication and erection of aluminum bridges and bridge components:

- Lighter weight and comparable strength enables the use of a higher ratio of live load to dead load, making the aluminum bridge girders and decks more efficient than steel or concrete components;
- Superior corrosion resistance eliminates the need to paint the aluminum components, except perhaps for aesthetic purposes, resulting in lower maintenance costs;
- Superior low-temperature toughness eliminates concerns about brittle fracture, even in the most severe Arctic weather;
- Ease of extrusion enables the design of more weight-efficient beam and component cross-sections, placing the metal where it is most needed within a structural shape or assembly, including providing for interior stiffeners and for joints; and
- The combination of light weight and ease of fabrication enables the entire aluminum structure or major portions of it to be pre-fabricated, carried to the site, and erected quickly with minimum interruption in the flow of traffic and thus less inconvenience to drivers.

It is appropriate to note, however, that there are some offsets to the advantages for aluminum that have deterred its broader usage in bridges, especially within the USA.

The most important of these is the higher initial cost (first cost) of aluminum bridge components over comparable steel and/or concrete components, which may depending upon design range from 25-75%. While the lower maintenance costs of aluminum bridge components over the life of the structure (principally, the lack of need for periodic painting) result in a lower total cost over the entire life of a bridge (life-cycle cost), the usual reliance upon separate new construction and maintenance budgets in most federal, state, and local highway districts has precluded the acceptance of the higher initial fabrication and erection costs in the USA. Greater use of aluminum for bridge structures has been made overseas (1-3) as will be described later, where such decisions are typically handled in a more centralized fashion.

Another factor limiting the use of aluminum for bridges has been the lack of general knowledge of the properties and design rules for aluminum in structural applications by many engineers and, as a result, their unwillingness to break away from the familiar patterns of use of steel and/or concrete structures. This is despite the availability for many years of the Aluminum Design Manual (1) produced by the aluminum Association, and aluminum's inclusion in most building codes.

In addition, colleges and universities focus primarily on steel and concrete when teaching structural engineering, with the result that few engineers come into the field with any knowledge of the use of aluminum in structural applications.

The net results is that despite the fact that aluminum makes up around 90% of the structural metal in aircraft and spacecraft, subject to very severe static and dynamic loadings, few civil and structural engineers in other industries know about the advantages of aluminum alloys.

There are some other factors that make designing aluminum bridge structures a bit different from designing steel structures, for example:
- Aluminum's lower modulus of elasticity (10 million psi [70 GPa] vs. 30 million psi [210 GPa] for steel).
- The fatigue strength of aluminum is about one-third that of steel.
- Aluminum's coefficient of expansion is about twice that of steel or concrete, so thermal stresses must be considered when aluminum components are fastened to these materials.

However such factors are readily handled by efficient design practices, for example, by using slightly deeper spans and thicker sections for aluminum than for steel. Even with such accommodations, aluminum structures, on average, weigh about one-half comparable steel structures.

Considerably more background and detail on aluminum structural design is included in References 2 through 9. The design specifications for aluminum structures (4) are included in the Aluminum Design Manual (1). Additional information of the design and application of

aluminum for a variety of structural applications including bridges is given in Reference 10.

The early applications of aluminum bridges in the USA

The practical use of aluminum in bridge applications can be traced to 1933 when the timber and steel floor system in the Smithfield Street Bridge in Pittsburgh, PA, was replaced by an aluminum deck (2). The change was made to significantly lighten the structure's deadload and, thereby, significantly increase its live-load-carrying capacity. The new deck was a riveted orthotropic deck, about 300 ft. (100 m) in length. The components were rolled 2014-T6 plate, the most widely used high-strength structural aluminum alloy at the time (though by today's standards it was not the best choice from the corrosion resistance standpoint).

This new aluminum deck structure enabled the bridge to carry the new electrified trolley cars being introduced at the time in the City of Pittsburgh. It carried two lanes of motor traffic and two tracks for trolleys moving both directions. The Smithfield Street Bridge became the major artery of the time carrying such traffic across the Monongehela River from the "Golden Triangle" to the South Side. The 1933 structure remained in service without problems for 34 years, until in 1967 the deck was upgraded with a new welded aluminum orthotropic deck, further increasing its ability to handle more and bigger trolleys and trucks. The new deck was of the orthotropic design, and the deck plate alloy was 5456-H321, a more corrosion-resistant alloy than the 2014-T6 used in the earlier deck. This plate was welded to 6062-T6 extrusions with 5556 filler wire; the extrusions were bolted to the bridge superstructure. This aluminum deck stayed in service without problems until 1993 when it was replaced by a steel deck (the decision being based upon short-term economics, not life-cycle cost).

The first all-aluminum bridge in the USA was constructed in 1946 for railroad traffic. One 100-ft (30.5 m) single-track span of a plate girder railroad bridge was constructed by Alcoa on a line serving their Massena smelter, probably as an illustration of the capability of aluminum in such applications. In this case, the girders were made of Alclad 2014-T6 plate, riveted with 2117-T4 rivets; the use of 2014 clad with relatively pure aluminum (1100) cladding was recognition of the lesser corrosion resistance of bare 2014 plate, as noted earlier.

The first all-aluminum highway bridge in the North America was erected in Arvida, Canada, over the Saguenay River in 1950 (Fig. 1). It was (and is today) a 290-ft (88 m) long, arch span bridge with multiple 20-ft (6 m) approach spans. It was erected by the Aluminum Company of Canada (Alcan), probably also as a working demonstration of aluminum's capability, and carried trucks with aluminum ores and products to and from various parts of the aluminum refining and smelting plants. It was then and remains today a very handsome bridge.

The 1950s and 1960s: Broader use of aluminum girder systems

In the period from about 1958 to about 1965, there was a national effort underway to upgrade the highway bridges across the USA and to find the most economical means of improving the safety of superhighways by incorporating cloverleaf intersections in them rather than dangerous crossroads. Aluminum alloys were among the materials of construction widely considered for these new or replacement bridges. In addition to its natural advantages, aluminum was seriously considered for bridges in the 1950s and 1960s in part because of the long lead times to obtain steel during that period. The interest was sufficient that, as we shall see, five significant new aluminum bridges were built in the USA over that seven-year period (1958-1967).

The first two of these were of relatively conventional built-up I-beam designs. A two-lane, four-span welded plate girder bridge was erected near Des Moines, IA, and a pair of two-lane, riveted plate girder bridges in Jericho, NY.

The Iowa bridge was a two-lane four-span bridge carrying 86th St. over I-80, and erected in 1958. The girders were of 5083-H113 aluminum plate welded with 5183 filler wire, with a concrete deck. These spans remained in service until about 1993, when they were removed because of an entirely new design of intersection being introduced at that location for which that bridge would no longer be needed. A thorough field and laboratory research program was conducted on the aluminum girder components as they were removed (3,11), and both tensile and fatigue tests of representative components of the girders were carried out. In every case, the test results showed that after about 40 years in service, the aluminum alloy members had tensile and fatigue properties comparable to those when the bridge was first erected and consistent with what would be expected in new structures today.

The twin Jericho, NY structures were two-lanes each, carrying I-495 traffic on the Jericho Turnpike, and were erected in 1960. The dual 77-ft (23.5-m) single span girders were fabricated of 6061-T6 plate with 2117-T4 riveted connections, and also had concrete decks. These spans were replaced in 1992 when the intersection was re-designed.

The last four aluminum bridge applications erected during that period were of the unique riveted, stiffened, triangular box beam girder concept referred to as the "Fairchild design" (12). This designation resulted from the fact that the design was conceived and put forth in the late 1950s by the Fairchild Kinetics Division of what was then called the Fairchild Engine and Airplane Company of Hagerstown, MD. Almost all commercial and military aircraft of the day were fabricated of high-strength aluminum alloys, and Fairchild engineers applied the current aircraft design concept of riveted, internally-stiffened sheet structures to bridge girder design. The design was also sometimes referred to as the "Unistress" design (13), because of the use of that term in a Kaiser Aluminum & Chemical Co. patent taken out about that time on one variation of it.

The cross-section of one of these bridges is illustrated in Fig. 2. It is a series of triangular box beams with common upper and lower flanges, plus end-frames. The result is a very stiff semi-monocoque design (2,12). In a monocoque structure, the skin absorbs all or most of the stresses to which the spans are subjected.

As part of the investigation of this innovative design, a full-scale 50-foot long bridge with a composite concrete deck was designed, fabricated, and tested by the Fritz Engineering Laboratory at Lehigh University (14). The advantages of the aluminum semi-monocoque design in providing lower dead load stresses (higher ratios of live load to dead load), lighter substructures, and reduced costs for transportation and erection were confirmed.

The first of the four bridges of the Fairchild or Unistress design was erected in 1961 in Petersburg, VA, carrying Route 36 over the Appomattox River. It was a single-span, two-lane bridge, with a concrete deck. The girder system was fabricated of 0.090-in (2.5 mm) 6061-T6 sheet. Like all three of the Fairchild designs it has remained in service for over 40 years.

As word of the opportunity to utilize a unique aluminum girder system to maximize the live loads of bridges became more widely known, construction of such bridges was also begun in Sykesville, MD and Amityville, NY.

The Sykesville Bypass Bridge, which carried MD Route 32 over the Patapsco River as well as the paralleling River Road and CSX Railroad (then the B&O), was the longest of this design ever built. The three nearly equal length spans total about 293 ft (almost 100 m). The MD State Highways Administration (SHA) engineers undertook the design of such a bridge for the planned new bypass of Route 32 around Sykesville, MD (15). Primarily because of (a) galvanic corrosion resulting from failure to maintain the isolation of the aluminum components and the steel bearings plus (b) an inadequate internal drainage system permitting water to lay inside the hollow sections, the Sykesville spans experienced galvanic and pitting corrosion; because of the high expense to repair it, the bridge was taken out of service in 2004 and replaced by an adjacent steel bridge.

In view of the unique nature of the bridge design and its use of aluminum components, the MD SHA, under the leadership of Rita Suffness, Architectural Historian and Cultural Resources Manager, recognized its historical significance and in coordination with the Maryland State Preservation Officer (MD SHPO), confirmed its place on the National Register of Historical Places in 1999. It has also been included in the MD SHA Historic Bridge Inventory, and in the Historic Bridges of Maryland (16).

The last two installations of the Fairchild design were in Amityville, NY, where a pair of three-lane, four-span bridges carry Route 110 over the Sunrise Highway. The Amityville spans, like the Sykesville spans, have deteriorated over the years; a renovation of the bridge, primarily to reduce the bearing stresses and assure isolation of bearing surfaces, has been proposed (17) and is being engineered by the New York State Department of Transportation.

Another innovative design of aluminum girder bridge seriously considered during this same period was conceived by Georgio Baroni for the Reynolds Metals Co. around 1958 (18). It employed a series of roll-formed inverted U-shaped cells placed longitudinally across the span, linked transversely, and integrated with the concrete deck to operate in semi-monocoque fashion. While believed to hold some promise at the time, and scheduled for erection in Alabama in 1960, the project was never completed.

A summary of the early aluminum usage in bridges in the USA is given in Table 1.

Table 1 – Early Use of Aluminum in Bridge Structures in North America

Location	Bridge Type	Use	Number Of Lanes	Span(s) m (ft)	Year Erected	Deck	Alloys Used
Pittsburgh, PA – Smithfield St.	Riveted Orthotropic Deck	Highway, Trolley	2 + 2 Tracks	Approx 100 (300)	1933	Aluminum Plate	2014-T6
Massena, NY – Grasse River	Riveted Plate Girder	Railroad	1 Track	30.5 (100)	1946	- - -	Alc 2014-T6 2117-T4 Rivets
Arvida, Canada – Saguenay River	Riveted Arch	Highway	2	5@6, 88, 5@6 (20,290,20)	1950	Concrete	2014-T6 Alc Plate, Extrusions 2117 rivets
Des Moines, IA – 86th St. over I-80	Welded Plate Girder	Highway	2	12,21,21,12 (41,69,69,41)	1958	Concrete	5083-H113
Jericho, NY I495 over Jericho Tpk	Riveted Plate Girder	Highway	4 (2 Bridges)	23.5 (77)	1960	Concrete	6061-T6
Petersburg, VA, Rte 36, Appomattox River	Bolted, Stiffened Triangular box Beam	Highway	2	29.5 (97)	1961	Concrete	6061-T6
Amityville, NY, Rte110 Sunrise Highway	Riveted Stiffened Triangular Box Beam	Highway	6 (2 Bridges)	18 (60)	1963	Concrete	6061-T6
Sykesville, MD, Rte 32 Patapsco River	Riveted Stiffened Triangular Box Beam	Highway	2	28,29,32 (93,94.106)	1963	Concrete	6061-T6
Pittsburgh, PA – Smithfield St.	New Welded Orthotropic Deck	Highway, Trolley	2 + 2 Tracks	Approx. 100 (300)	1967	Aluminum Plate	5456-H321

The use of aluminum bridge structures overseas

A summary of some of the earlier applications overseas is given in Table 2. There was considerable interest in what was going on in the USA, as illustrated by the attention to the USA applications described in the European press (examples: Ref. 12 and 13, articles from the French metallurgical publication, Revue De L'Aluminium).

As illustrated in Table 2, the three earliest European applications, from 1948-1950, were all in Great Britain, two of which were movable bascule bridges taking advantage of the lighter weight of aluminum spans (22,23), and the other a pedestrian bridge (24). Over the next ten years, six other aluminum structures were erected in Germany, Switzerland and England; four of the six were pedestrian bridges. Additional information on some of these applications is included in References 2 and 4.

While there was no widely publicized use of aluminum bridges in France until around 1968-1970, several applications during and after that period were reported in the principal French journal on aluminum, Revue De L'Aluminium (25-31). For example, it was reported that "the world's longest pedestrian bridge" at the time was erected in 1968/9 at Dunkerque (25) and, beginning in 1973, several bridge deck replacements of aluminum for steel/concrete decks were made to increase live load capacity of relatively old bridges (e.g., Montmerle, 1973; Groslee, 1976/7; and Chamalières, about 1978). The bridge at Chamalières (30) is of special interest as it also employed an aluminum girder system in its upgrade to permit widening the bridge from two to four lanes.

Table 2 – Early Use of Aluminum in Bridges Overseas

Location	Bridge Type Connection	Use	Number Of Lanes	Span(s) m (ft)	Year Erected	Deck	Alloys Used
Hendon Dock, England	Riveted Double Leaf Bascule	Highway, Rail	1 + 1 Track	37 (121)	1948	Aluminum Plate	2014-T6 6151-T6
Tummel River Scotland	Riveted Truss	Pedestrian	- - -	21,52,21 (69,172,69)	1950	Aluminum Sheet	6151-T6
Aberdeen, Scotland	Riveted Double Leaf Bascule	Highway, Rail	1 + 1 Track	30.5 (100)	1953	Aluminum Sheet & wood	2014-T6 6151-T6
Dusseldorf, Germany	Twin Web Plate, Arched Ribs	Pedestrian	- - -	55 (180)	1953	- - -	- - -
Lunen, Germany	Riveted Warren Truss	Highway	1	44 (145)	1956	Aluminum Shapes	6351-T6
Lucerne, Switzerland (two bridges)	Suspension Stiffened Girder	Pedestrian & Cattle	- - -	20 (65) 34 (112)	1956	Timber	5052
Rogerstone South Wales	Welded W Truss, Thru Girder	Pedestrian	- - -	18 (60)	1957	Corrugated Aluminum Sheet	6351-T6
Monmouth-Shire, England	Welded	Pedestrian	- - -	18 (60)	1957	Corrugated Aluminum Sheet	6351-T6
Banbury, England	Riveted Bascule	Highway	1	3 (9.5)	1959	Corrugated Al Sheet	6351-T6
Gloucester, England	Riveted Bascule	Highway	1	12 (40.5)	1962	Extruded Al Shapes	6351-T6

Aluminum bridge applications today

In the mid-1990's Reynolds Metals (now a part of Alcoa, Inc.) developed several aluminum bridge deck designs. The first was developed specifically for the 320 ft long, 12'-6" wide historic Corbin suspension bridge over the Juniata River near Huntingdon, PA (Fig. 3). The steel bridge deck had deteriorated, limiting the live load to 7 tons. The

aluminum replacement deck was an approximately 5 in. (130 mm) deep 6063-T6 multiple hollow extrusion that was welded on its top flange only. The deck extrusion was oriented transversely to traffic and supported by 10 in. (250 mm) deep 6061-T6 aluminum extruded I beams oriented parallel to traffic. By reducing the dead load, the new deck permitted the live load rating of the bridge to be increased to 22 tons.

The second Reynolds bridge deck was used on a US Route 58 bridge over the Little Buffalo Creek near Clarksville, VA. The bridge was 54'-9 ¾" long and 32'-0 wide. This deck was made of 12" deep 6063-T6 extrusions welded on both the top flange and the bottom flange from one side with a removable backing. The extrusions were oriented parallel to traffic and attached to the four longitudinal steel bridge beams.

An issue with aluminum bridge decks encountered by Reynolds and others is the need to provide a surface on the aluminum that affords slip resistance to traffic. The Reynolds decks used a 3/8" thick epoxy with embedded aggregate, similar to that used occasionally on concrete decks.

There has been in recent years relatively greater use of aluminum bridge decks overseas, primarily to replace or update decks in older bridges where capability to carry greater live load is an objective. In Sweden, for example, the Svensson/Petersen design (1,3,32) using hollow stiffened 6063-T5 or T6 extrusions (Fig. 4) has been in use for about 20 years in more than 30 replacement decks around Stockholm alone; an example is illustrated in Figs. 5. In this case of a bridge providing access to a major throughway near Stockholm in which, in order to minimize traffic interruption in a major artery, the deck was replaced overnight.

In 1995, an all-aluminum installation in Norway at Forsmo (3) used an aluminum deck combined with aluminum girders. This provided a quickly assembled portable structure that could be carried on a truck to the site (Fig. 6) and dropped into place in a single crane operation (Fig. 7). The cross-section of the Foresmo bridge is shown in Figure 8, and the completed bridge is shown in Fig. 9.

The most recent bridge deck installation in the U.S. was that in Clark County, KY (33), where a fast replacement was needed for a rural road carrying school and hospital traffic. The deck was made with hollow, integrally stiffened 6063-T6 shapes pre-fabricated off-site and placed in position in just a few hours, minimizing the disruption in traffic.

Life-Cycle Cost Analysis is Key

For today's application of aluminum alloys to bridges and bridge decks, internally stiffened hollow extruded panels, similar to the U.S. or Swedish designs, are of most interest, and assessment of their economic value must be based upon total life cycle cost, not initial erection cost alone.

As an example, using hypothetical figures to avoid variations resulting from site-related variables, let us assume

that a steel bridge deck can be built for $75/sqft, and a 300-ft long, two lane (20 ft) wide bridge is to be built: the nominal cost of the deck in steel would be $450K. The cost of an aluminum deck would be $125/sqft; that cost would be $750K. However the steel deck will have to be painted every ten years of a 50-year assumed life, and the cost of each re-painting is estimated to be about 1/3 the cost of original construction because of the considerable environmental requirements. Therefore the cost of the steel deck over its 50-year life would be $450K + 4x$150K or $1,050K. The aluminum deck never has to be repainted, so the life-cycle cost remains $750K.

There are the following specific added benefits provided by the fact that the aluminum bridge deck may be pre-fabricated off-site, in whole or two or three sections, and transported to the site for erection. Therefore the bridge does not have to be closed until it is time to erect the new span, greatly minimizing the closure time for the bridge and the disruption of traffic for drivers. It is difficult to place a monetary value of such savings, but they are considerable in the public mind.

Aluminum alloys recommended for bridge decks

A great number of aluminum alloys might be chosen for bridge and bridge deck construction (5), but those most highly recommended and used currently because of their superior combination of strength, corrosion resistant, and overall ease of fabrication are illustrated in Table 3. In general, alloys of the 5xxx series are used for the plate components, and the 6xxx alloys are used for the extruded shapes. Alloy 6063 is a particular favorite for the latter if complex and/or hollow sections are required.

It is also appropriate to note at this stage that the Fairchild design of aluminum bridge discussed earlier (Sections 3.0 and 4.0) would not be considered practical and cost effective today. The complex buildup of sheet and extrusion components and its riveted construction is very labor intensive and, hence, very expensive compared to other designs of aluminum structures. Current design practice is for the use of structural aluminum decks in combination with steel or reinforced concrete girders, where the aluminum decks are made up of long lengths of aluminum plates and/or extrusions requiring only minimal assembly.

Conclusions

Aluminum alloys have much to offer for bridge and bridge deck applications, and continue to be used, primarily overseas, where their light weight, high strength-to-weight ratio, and excellent corrosion resistance satisfy service requirements.

Aluminum alloys have been used in U.S. bridge structures since 1933, when the first aluminum bridge deck was used to replace an earlier steel and wood deck on Pittsburgh's Smithfield Street Bridge in order to increase its live-load carrying capacity. Aluminum girders and bridge decks have been used in about 10 other bridges in the USA, the most recent recognized here in 1997. While still not considered a standard for bridge structures in the USA, they are more widely used overseas, as for example around Stockholm, where more that thirty bridges have been rehabilitated with aluminum decks.

With proper design and maintenance, aluminum girders and decks may provide lower life cycle costs and as long or longer lives than steel and/or concrete alternatives.

Acknowledgements

The authors wish to acknowledge the Aluminum Association, Inc., notably the Engineering & Design Task Force, J. Randolph Kissell, Task Force Administrator, as the basis of much of the information contained herein.

References

1. *Aluminum Design Manual 2005*, The Aluminum Association, Inc., Washington, DC, 2005.

2. *Aluminum's Potential: Bridge Construction, The Proceedings of the Bridge Session of the 1994 Alumitech Conference,"* October 26-29, 1994, Atlanta, GA., The Aluminum Association, Inc., Washington, DC, 1995.

3. *Aluminum Alloys for Bridges: Extending the Life of U.S. Bridges*, The Aluminum Association, Inc., Washington, DC, 1996.

4. *Aluminum in Bridge Construction: Perspective on Design, Construction, and Performance*, The Aluminum Association, Washington, DC, November, 1997.
   - "Forsmo Bridge – The First Aluminium Road Bridge in Normay" by I. Kvale, J Vallestad, and K. Solaas
   - "Aluminum Bridges and Bridge Decks for Lowest Life-Cycle Costs" by J.J. Ahlskog
   - "Aluminum Bridges – Comparative Design Cases" by Dimitris Kosteas
   - Load Tests of and Aluminum Girder Highway Bridge and Girder Components" by R.E. Abendroth and W.W. Sanders

5. *Specifications for Aluminum Structures*, The Aluminum Association, Inc., Washington, DC, 2005 (published periodically).

6. M.L. Sharp, *Behavior and Design of Aluminum Structures*, McGraw-Hill, Inc., New York, 1993.

7. M.L. Sharp, G.E. Nordmark, and C.C. Menzemer, *Fatigue Design of Aluminum Components and Structures*, McGraw-Hill, Inc., New York, 1996.

8. Bucci, G. Nordmark, and E.A. Starke, Jr., *Selecting Aluminum Alloys to Resist Failure by Fracture Mechanisms*, ASM Handbook, Volume 19, Fatigue and Fracture, ASM International, Materials Park, OH, 1996, pp. 771-812.

9. J. R. Kissell and R.L. Ferry, *Aluminum Structures, A Guide to Their Specifications and Design* Second Edition, John Wiley & Sons, New York, 2002.

10. D.G. Altenpohl, *Aluminum – Technology, Applications, and Environment*, Aluminum Association, Inc., and TMS, Inc., Washington, DC and Warrendale, PA, 1998 (second Printing, 1999).

11. "A Continuous Span Aluminum Girder Concrete Deck Bridge," Final Report, Part 1: Field Test Performance and Evaluation, by R.E. Abendroth, W.W. Sanders, and V. Mahadevan, July, 1996; and Part 2: Fatigue Tests of Aluminum Girders, by R.E. Abendroth, W.W. Sanders, and S. Hansz, May, 1997, Center for Transportation Research and Education, Iowa State University, Ames, Iowa.

12. Le Pont Fairchild Sera-T-Il Le Prototype Des 70,000 Ponts Du Reseau Routier Americain," Revue de L'Aluminium No. 266, June, 1959.

13. "Le premier pont routier "Unistress," Revue de L'Aluminium No. 308, June, 1963.

14. H.L. Mindlin and S.J. Errara, "Test of a Composite Aluminum and Concrete Highway Bridge," Report 275.1, Fritz Engineering Laboratory, Lehigh University, Bethlehem, PA, 1959.

15. MD SHA REPORT MD-100, *Stress Analysis 105.17 Foot Semi-Monocoque Aluminum Bridge Structure for Sykesville By-Pass, Maryland*, by V.J. Pernetti, Maryland State Highways Administration, Baltimore, Maryland, December 1960.

16. Dixie Legler and Carol M. Highsmith, *Historic Bridges of Maryland*, Maryland Historical Trust Press, 2002.

17. Osman Hag-Elsafi and Sreenivas Alampali, "Cost-Effective Rehabilitation of Two Aluminum Bridges on Long Island, New York," Practice Periodical on Structural Design and Construction, August, 2002, pp. 11-17.

18. "Le pont Routier Reynolds-Baroni," (The Reynolds-Baroni Highway Bridge) Maurice Victor, Revue de L'Aluminium, No. 271, December, 1959, pp. 1321-1323.

19. *"Historical Documentation of Bridge No. 13046, Sykesville, Maryland, MD 32 over Patapsco River, CSX Railroad, and River Road, an Aluminum Girder Bridge,"* Report of the Aluminum Association Bridge Team to MD SHA, March, 2003.

20. "Sverdrup Engineering Report, "Inspection and rating of State, County, and Local Jurisdiction Bridges – Contract BCS 36-18C, Task 7 – In-depth Inspection, Analysis and Rating Bridge No. 13046 – MD Route 32 over River Road, South Branch Patapsco River, and B&O R.R.," transmitted by letter Tuhin K. Basu to Geoffrey V. Kolberg, January 19, 1989.

21. *"Report on the Condition of Bridge No. 13046, Sykesville, Maryland, MD 32 over Patapsco River, CSX Railroad, and River Road, an Aluminum Girder Bridge,"* Report of the Aluminum Association Bridge Team to MD SHA, March, 2003.

22. "Le pont du Hendon Dock a Sunderland," (The Sunderland Bridge at Herndon Dock), Jean Romeyer, Revue de L'Aluminium, No. 155, May, 1949, pp. 158-162.

23. "Un noveau pont Basculant entierement en alliage leger," Jean Reinhold, Revue de L'Aluminium, No. 208, March, 1954, pp. 98-100.

24. "un pont en Duralumin de 95 metres de portee sur la Riviere Tummel," Maurice Victor, Revue de L'Aluminium, Maurice Victor, (No. and Date lost), pp. 56-57.

25. "La plus grande passerelle du monde en aluminium: 83 metres," (The Longest Pedestrian Bridge in the World: 83 meters, about 255 ft.), Revue de L'Aluminium, No. 381, January, 1970, pp. 1-8.

26. "le pont de Montmerle," (The Bridge at Montmerle), Revue de L'Aluminium, April, 1973, pp. 245-248.

27. Le premier pont tous tonnages en aluminium," (The First Bridge Deck Made Entirely of Aluminum), Revue de L'Aluminium, No. 447, January, 1976, p. 41.

28. "Le pont de Groslee en mai 1977," A. Brisseaux, Revue de L'Aluminium, No. 464, June, 1977, pp. pp. 386-387.

29. "L'aluminium trait d'union entre le Bugey et le Dauphine," Revue de L'Aluminium, No. 468, December, 1977.

30. "L'aluminium sur les routes de la Haute-Loire," (Use of Aluminum on the Roads of the High Loire), A. Brisseaux, Revue de L'Aluminium, No. 478, November, 1978, pp. 496-500.

31. "le passerelle emn aluminium de Villepinte," (A Pedestrian Bridge in Aluminum from Vilepinte to the airport Roissey-Charles De Gaulle), A. Brisseaux, Revue de L'Aluminium, June, 1980, pp. 279-282.

32. Svensson, Jars and Pettersen, Lars, "Aluminum Extrusion Bridge Rehabilitation System," *Bridge Maintenance – Inspection, Maintenance, Assessment, and Repair*, Elsevier, London, 1990.

33. Press Release "Aluminum Deck on a Kentucky Bridge," Issam Harik, University of Kentucky, July 18, 2006

Listing of Figure Captions

Figure 1 - All-aluminum arch span over Saguenay River in Arvida, Canada, erected 1950, still in service today

Figure 2 - Drawing of cross-section of girder system of Route 32 Sykesville Bypass Bridge, As-built drawing, MD SHA Archives

Figure 3 - Aluminum hollow extruded 6063-T6 shape used in Reynolds design of bridge deck

Figure 4 - The historic Corbin Bridge near Huntington, PA, where an innovative aluminum replacement deck was used to increase live-load capability by 300% in 1996

Figure 5 - Representative cross-sections of 6063-T6 extruded shapes making up Swedish bridge deck system

Figure 6 - Stockholm highway bridge for which an aluminum deck was installed overnight to minimize traffic interruption on a major artery

Figure 7 - View of the Forsmo, Norway all-aluminum bridge during transportation to the site, illustrating one important advantages of lightweight aluminum, bridge construction

Figure 8 - View of the Forsmo, Norway all-aluminum bridge during erection, illustrating another important advantages of lightweight aluminum, bridge construction

Figure 9 - Cross-section illustrating design of all aluminum girder system of Forsmo Bridge

Figure 10 - View of Forsmo Bridge completed for service

Table 3 – Minimum (design) properties of some aluminum alloys for bridge components (Ref: The Aluminum Design Manual[1,2]

Alloy	Temper	Product	Thickness Range	Tension Ultimate Strength	Tension Yield Strength	Compression Yield Strength	Shear Ultimate Strength	Shear Yield Strength	Compressive Modulus of Elasticity Y(3)
Metric/SI units			mm	MPa	MPa	MPa	MPa	MPa	GPa
5083	O	Sheet & Plate	1.20-6.30	275	125	125	170	70	71.7
5083	H116	Sheet & Plate	4.00-40.00	305	215	180	180	125	71.7
5083	H321	Sheet & Plate	4.00-40.00	305	215	180	180	125	71.7
5086	O	Sheet & Plate	0.50-50.00	240	95	95	145	55	71.7
5086	H32	Sheet & Plate	All	275	195	180	165	110	71.7
5086	H34	Sheet & Plate	All	300	235	220	180	140	71.7
5086	H116	Sheet & Plate	All	275	195	180	165	110	71.7
5454	O	Sheet & Plate	0.50-80.00	215	85	85	130	48	71.7
5454	H32	Sheet & Plate	0.50-50.00	250	180	165	145	105	71.7
5454	H34	Sheet & Plate	0.50-25.00	270	200	185	160	115	71.7
6061	T6, T651X	Extruded Shapes	All	260	240	240	165	140	69.6
6063	T5	Extruded Shapes	Up thru 12.50	150	110	110	90	62	69.6
6063	T5	Extruded Shapes	12.50-25.00	145	105	105	85	59.0	69.6
6063	T6	Extruded Shapes	All	205	170	170	130	95	69.6
Engineering units			in.	ksi	ksi	ksi	ksi	ksi	103 ksi
5083	O	Sheet & Plate	0.051-1.500	40	18	18	25	10	10.4
5083	H116	Sheet & Plate	0.188-1.500	44	31	26	26	18	10.4
5083	H321	Sheet & Plate	0.188-1.500	44	31	26	26	18	10.4
5086	O	Sheet & Plate	0.020-2.000	35	14	14	21	8	10.4
5086	H32	Sheet & Plate	All	40	28	26	24	16	10.4
5086	H34	Sheet & Plate	All	44	34	32	26	20	10.4
5086	H116	Sheet & Plate	All	40	28	26	24	16	10.4
5454	O	Sheet & Plate	0.020-3.000	31	12	12	19	7	10.4
5454	H32	Sheet & Plate	0.020-2.000	36	26	24	21	15	10.4
5454	H34	Sheet & Plate	0.020-1.000	39	29	27	23	17	10.4
6061	T6, T651X	Extruded Shapes	All	38	35	35	24	20	10.1
6063	T5	Extruded Shapes	Up thru0.500	22	16	16	13	9	10.1
6063	T5	Extruded Shapes	0.501-1.000	21	15	15	12	8.5	10.1
6063	T6	Extruded Shapes	All	30	25	25	19	14	10.1
Footnotes		1 - Reference: Aluminum Design Manual 2000, The Aluminum Association							
		2 - For tensile yield strengths, offset = 0.2%							
		3 - Typical values; for deflection calculations, average modulus of elasticity is used, which is 0.1 ksi or 0.7 Gpa lower than values in this column							

Fig 1

Fig 2

Fig 3

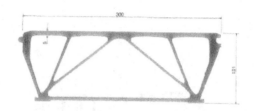

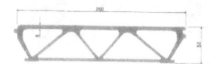

Fig 4

Fig 5

Fig 6

STUDY ON EVOLUTION BEHAVIORS OF BUBBLE IN ALUMINUM MELT

H.J. LUO, G.C. YAO, Y.H. LIU, X.M. ZHANG

School of Materials and Metallurgy, Northeastern University, Shenyang, 110004, China

Keywords: foaming, aluminum foam, bubble, growth, plate

Abstract

The nucleating method of bubble in aluminum melt was studied in this paper. The driving force of bubble growth and its evolution behavior were also analyzed. The results show that the bubble formation in aluminum melt belongs to heterogeneity nucleation and the existence of solid particles provides necessary condition. The bubble growth achieves in the stage of constant temperature foaming mostly. The driving force of bubble growth comes from releasing hydrogen of TiH_2. Two periods can be divided in the stage of bubble growth. The major growth manner of bubbles is expansion in prior period and coalescence in final period. Bigger diameter bubbles grow up prior to smaller ones.

Introduction

Aluminum foam is a composite material consisting of metallic matrix and gas bubbles. It can be widely used in transportation, packaging, aerospace, electrommunication, etc., due to its characteristics of high specific strength, fire-resistance, sound and energy absorption, sound isolation, impact absorption and interception of electric wave [1,2]. The melt foaming method is a main manner to fabricate aluminum foam, and the method can be divided into such routes as endogenous gas foaming and injecting gas foaming further from sources of foaming gas [3~7]. The endogenous gas foaming is a technique that solid foaming agent is added into molten metal, and gas decomposed by the foaming agent makes the melt expanded. However, The bubbling air is introduced into molten metal when using injecting gas foaming to fabricate aluminum foam. The interaction among gaseous phase, fluid phase and solid phase is controlled, and bubbles float to the top surface of the melt to form closed cell foam. Much effort has been directed towards the process by such ways as bubble nucleation, evolution, stabilization due to the generation and change of bubble directly affecting the structure of foam body [8~14]. The objective of this paper is to fabricate aluminum foam through endogenous gas foaming method with emphasis on evolution manners of bubbles. The process of using this method to prepare aluminum foam is below: increasing viscosity agent, such as Ca, was added into the molten aluminum or its alloys first, then foaming agent, such as TiH_2, was also put into the melt. The melt including foaming agent was put in a constant temperature stove to wait for foaming after being stirred uniformly. The quadrate body with cellular structure could be obtained after cooling, and it was sawn in order to gain aluminum foam plate, shown in figure 1.

Bubble Nucleation

The generation of bubble in molten aluminum containing TiH_2, in addition to relating to the gas decomposed by TiH_2, is also affected and restricted by the melt properties. Supposing there is a bubble suspending 'h' cm below the melt surface, the pressure inside the bubble exerted by hydrogen is p_h, and an approximate equality is given by

$$p_h = p_a + p_l + p_s \tag{1}$$

$$p_h = p_a + rgh + 2s/r \tag{2}$$

where p_a is atmosphere pressure, p_l is the static pressure exerting the bubble surface, p_s is additional pressure generated by interfacial tension, ρ is the melt density, g is the gravitational acceleration, σ is the interfacial tension of the bubble, and r is the bubble radius. Assuming the purity of aluminum is 99.9%, and its density and interfacial tension are 2.45 g/cm³ and 0.91 N/m respectively, the pressure that a initial bubble has to overcome in atmosphere at 660°C can be calculated. Namely, for a bubble with radius being 1×10^{-5}cm, and depth below the melt surface about 5cm, the pressure endured by the bubble, p_h, is 1.83×10^7Pa.

The viscosity of molten aluminum or its alloy increases along with the addition of Ca, and their interfacial tension is larger than that of pure aluminum liquid. Therefore, the nucleating pressure of bubble coming from the mixture is also bigger than above calculated value, and homogeneous nucleation in the mixture is nearly impossible. It can be determined that the bubble nucleation belongs to heterogeneous nucleation. Figure 2 shows a piece of X-ray diffraction spectra of ternary alloy with addition 3% Ca to Al-Si eutectic alloy and then stirring 10 minutes. The particulate matter ($CaAl_2Si_2$) can be found existing in the ternary alloy. Meanwhile, TiH_2 itself can also be regarded as solid particles. The existence of these particles reduces the nucleating energy of bubble and makes heterogeneous nucleation in molten aluminum possible [7,8].

Fig.1 aluminum foam plate

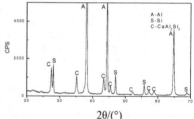

Fig.2 X-ray diffraction spectra
of Al-12%Si-3%Ca alloy

74

Bubble Growth

Driving Force of Bubble Growth

Equation (1) and (2) show that the inner pressure (p_h) is equal to outside pressure of a bubble (p_a, p_l, p_s) in balance, and p_h must be greater than the outside pressure when the bubble grows, which is

$$p_h > p_a + p_l + p_s \tag{3}$$

If the state of gas in the bubble is expressed by prefect gas equation, the following equation can be shown,

$$P_h V = nRT \tag{4}$$

where V is the bubble volume, n is the gas mole number in the bubble, R is the gas constant, T is the absolute temperature, and the bubble volume can also be expressed as follow:

$$V = \tfrac{4}{3} p r^3 \tag{5}$$

If substituting equation (2) and (5) for equation (4), and doing differential calculation against time, the following relationship can be obtained.

$$\frac{dn}{dt} = \frac{4pr}{RT}[\, r\, (\, \text{Pa} + r\,\text{gh}) + \frac{4}{3} s\,]\frac{dr}{dt} \tag{6}$$

The equation (6) indicates that a functional relation is established between growing rate of a bubble (dr/dt) and decomposing rate of foaming agent (dn/dt). Because the atmosphere pressure (p_a), static pressure (p_l), interfacial tension (σ) and the melt temperature (T) are constant, the bubble radius will enlarge according to increasing of hydrogen amount in the bubble, namely the decomposing rate of foaming agent is the driving force of bubble growth.

Manner of Bubble Growth

The manner of bubble growth in molten aluminum may be divided into two conditions according to whether stirring is existent. Although gas bubbles grow in the stirring stage, tremendous stirring force and centrifugal force throw them against the inside wall of crucible, and they are entrapped into spiral vortex along with writhing of melt, resulting in colliding with stirring paddle under the spiral vortex and being smashed. The size and distribution of bubbles tend to be uniform in above reciprocating movement. It can be concluded that the bubble growth is in a restrained state in the condition of stirring and its size remains within some range, and this is the reason why the change of melt volume is relatively small in the stirring period.

The aluminum melt comes into the constant temperature foaming stage after moving away the paddle, and bubble growth will be completed in this stage. The bubble growth still obeys equation (3), namely inner pressure is greater than the sum of outside pressure. The atmosphere pressure is a certain value through analyzing equation (1) and (2), and the static pressure endured by bubbles is associated with their position in the melt, but the value of static pressure is relatively small. Accordingly, additional pressure plays a key role in the process of bubble growth. Considering the interfacial tension (σ) is fixed in the factors that modify the additional pressure, the bubble radius is the most key factor, which is directly related to whether the bubble grows and how many size it can turn into.

There are two ways for existence of TiH_2 in molten aluminum in the constant temperature foaming stage. Adhering to the generated bubble is one way, and entrapping in the melt around bubbles is another. The TiH_2 particles adhering to bubbles tend to release gas into the inside of bubbles due to the huge differential value of additional pressure ($\triangle 2s/r$), and driving bubble expansion further. It is difficult for the TiH_2 particles entrapping in the melt to release gas or form new bubble because forming a new bubble with radius nearly zero must overcome huge additional pressure. However, hydrogen released by these TiH_2 particles increases the hydrogenous concentration around bubbles, and supersaturated hydrogen can also diffuse itself into bubbles.

It can be seen from the above analysis that big bubbles grow up prior to small ones in the constant temperature foaming period. Assuming there is a new bubble generated in the liquid with regard to heterogeneous nucleation, it is difficult for it to grow due to the relatively still state of the melt, seldom chances of the new bubble adsorbing other solid particles as well as its minute radius. Figure 3 gives the longitudinal cross-section SEM micrograph of aluminum foam after foaming in constant temperature, where (a) is an amplified micrograph of a group of cells, and (b) is cross section of a single cell wall. As seen in (a) and (b), there are lots of small bubbles distributed both in inside surface of cells and in cross section of cell wall. Although it can't be determined which stage these small bubbles occur, the limited cell size of these bubbles is confirmed. The bigger bubbles in molten aluminum grow up sufficiently and form the main structure of aluminum foam.

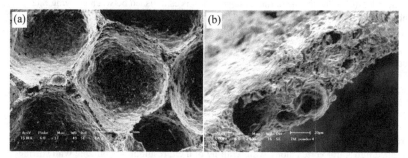

Fig.3　SEM micrograph of bubbles after foaming in constant temperature, (a) an amplified micrograph of a group of cells；(b) the cross section of a single cell wall

Evolution of Bubble

The bubble growth may also be divided into two processes in constant temperature foaming stage. The bubbles grow by means of expansion in early stage, and the interaction and effect among bubbles are small. When bubbles expand to some extent and form common cell wall among bubbles, a bubble growth is not only restricted by surrounding bubbles but also affects the final foam structure. The SEM micrographs of cross section with common cell wall between or among adjacent bubbles are shown in figure 4, where (a) is a common cell wall between two bubbles, and (b) is a common cell wall (Plateau border) among three bubbles.

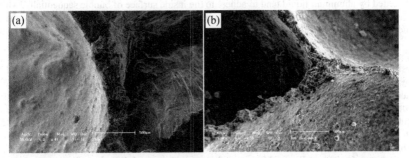

Fig.4　SEM micrograph of cell wall, (a) the common cell wall between two bubbles, (b) the common cell wall (Plateau border) among three bubbles

Besides affected by the driving force coming fromTiH$_2$ releasing hydrogen, the following actions also influence the bubble evolution significantly. One action comes from bubble growth, upward floating trend and interfacial tension, and the melt around bubbles is extruded, resulting in downward flowage of aluminum liquid and forming so-called drainage phenomenon. The cell wall becomes thinner and thinner along with proceeding of drainage. Another action is adsorption generated by bubble itself. A lot of solid particles are attached to the surface of bubble by adsorption action. The solid particles adhering to the surface of bubble can restrict cell wall to thin, because they should be extruded from the bubble surface as cell membrane becomes thin enough [11]. Though the solid particles adhering to the surface of bubble can baffle the proceeding of drainage, the fracture of common cell wall among bubbles reduces surface energy and is of spontaneous trend. Therefore, when the capacity that TiH$_2$ releases hydrogen is surplus, the cell wall at critical state will rupture and coalescence among bubbles will be inevitable. Figure 5 is the optical photographs of coalescence between or among bubbles (see (a) and (b)).

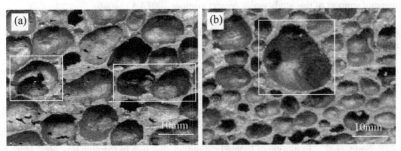

Fig.5　Optical photograph after merging between or among bubbles, (a) the connected cells after merging between bubbles, (b) a big cell after merging among bubbles

It can be observed in experiment and test that the solid particles adhering to cell wall help to stabilize bubbles [15], and the stabilization is concerned with wettability between these particles and aluminum liquid. Bubbles can easily adsorb the particles that can't be wetted by the liquid. Despite bubbles adsorbing particles wetted by aluminum liquid partly, these particles are gradually squashed into cell wall in the course of the cell wall being thin. Whereas, the particles not wetted by aluminum liquid tend to adhere to the inside surface of bubble sequentially, which is the side near gas phase in bubble. It is also necessary to explore the definite mechanism that solid particles stabilize bubbles in melt.

The bubble connected locally can't maintain its existing shape due to the action of interfacial tension, and ruptured membranous cell wall will vanish little by little and two connected bubbles may change into a long shaped or elliptical bubble. The bubble growth is saltatory by means of coalescence in final stage. If the coalescent behaviors happen uniformly and spread all over bubbles, the obtained structure of aluminum foam is still uniform to some extent, as shown in figure 5 (a), otherwise, the uniformity of foam structure may be damaged, as shown in figure 5 (b). It should be noted that the coalescence occurring among bubbles continuously may lead to form a cavity in middle of the foam body and even make the whole foam body collapsed, if TiH_2 foaming capacity is superfluous. Moreover, the precipitation and diffusion of hydrogen can make bubbles a little big in cooling of foam body, because hydrogenous solubility in solid-state aluminum is far lower than that in liquid-state aluminum.

Conclusion

The bubble nucleation in aluminum melt belongs to heterogeneous nucleation and the existence of solid particles provides a precondition for heterogeneous nucleation of bubble. The bubble growth is mainly finished in constant temperature foaming stage, and the hydrogen released by undecomposed TiH_2 is the driving force of bubble growth. The bubble growth can be divided into two processes in this stage. The bubbles grow by means of expansion in early stage and by means of coalescence in final stage. The bubbles with bigger diameter are prior to grow and form the main structure of aluminum foam. A massive aluminum foam plate can be fabricated by using the above method.

Acknowledgements

The authors would like to thank the Ministry of Education of the People Republic of China for the financial support of this study under project number 2002AA334060.

References

[1] J. Baumeister, J. Banhart, and M. Weter, "Aluminium foams for transport industry," *Materials & Design*, 18(4/6)(1997), 217-220.

[2] F. Yi, et al., "Strain rate effects on the compressive property and the energy-absorbing capacity of aluminum alloy foams," *Materials Characterization*, 2001, no.47: 417– 422

[3] K. Wu, et al., "Average foaming life and foaming intensity of foaming process originated from reaction," *The Chinese Journal of Nonferrous Metals*, 10(6)(2000), 909-913 (In Chinese).

[4] Z. L. Song, et al., "Evolution of foamed aluminum structure in foaming process," *Materials Science and Engineering*, 2001, no.A298: 137–143.

[5] J. T. Beals and M. S. Thompson, "Density gradient effects on aluminum foam compression behavior," *Journal of Materials Science*, 1997, no.32: 3595-3600.

[6] D. Q. Wang, Z. Y. Shi, "Effect of ceramic particles on cell size and wall thickness of aluminum foam," *Materials Science and Engineering*, 2003, no.A361: 45–49.

[7] C. C. Yang , H. Nakae, "The effects of viscosity and cooling conditions on the foamability of aluminum alloy," *Journal of Materials Processing Technology*, 2003, no.141: 202–206.

[8] S. V. Gniloskurenko, et al., "Theory of initial microcavity growth in a liquid metal around a gas-releasing particle. П. bubble initiation conditions and growth kinetics," *Powder Metallurgy and Metal Ceramics*, 41(1-2)(2002), 90-96.

[9] S. W. Ip, Y. Wang and J. M. Toguri, "Aluminum foam stabilization by solid particles," *Canadian Metallurgical Quarterly*, 38(1) (1999), 81-92.

[10] B. L. Mordike and P. Lukáč, "Interfaces in magnesium-based composites," *Surface and Interface analysis*, 2001, no.31: 682–691.

[11] C. Körner, et al., "Endogenous particles stabilization during magnesium integral foam production," *Advanced Engineering Materials*, 6(6)(2004), 385-390.

[12] H. P. Degischer, "Innovative light metals: metal matrix composites and foamed aluminum," *Materials & Design*, 18(4/6)(1997), 221-226.

[13] G. Kaptay, "Interfacial criteria for stabilization of liquid foams by solid particles," *Colloids and Surfaces A: Physicochem. Eng. Aspects*, 2004, no.230: 67–80.

[14] V. Gergely, T. W. Clyne, "Drainage in standing liquid metal foams: modelling and experimental observations," *Acta Materialia*, 2004, no.52: 3047-3058.

[15] H. J. Luo, et al., "Fabrication aluminum foam with fly ash as viscosifier," *Journal of Northeastern University (Natural Science)*, 26(3)(2005), 274-277(In Chinese).

Aluminum Alloys for Transportation, Packaging, Aeropsace, and Other Applications
Edited by Subodh K. Das, Weimin Yin
TMS (The Minerals, Metals & Materials Society), 2007

Sound absorption property of closed-cell aluminum foam

Haijun Yu, Guangchun Yao, Yihan Liu, Hongjie Luo, Guojun Yang

School of Materials and Metallurgy, Northeastern University. P.O. Box 117[#], Shenyang, Liaoning, 110004, P.R. China

Keywords: closed-cell aluminum foam, sound absorption, mechanism, thickness, porosity

Abstract

Closed-cell aluminum foams of different porosities and different thicknesses were prepared in this test; the sound absorption property was tested by utilizing standing wave tube, the sound absorption mechanism of which was studied, and the effect of porosity and thickness on its sound absorption property was also researched. It is found that the sound absorption property of closed-cell aluminum foam mainly depends on the Helmholtz resonators, the micropores and cracks of material. The porosity and thickness obviously affect the sound absorption property of closed-cell aluminum foam: the sound absorption coefficient increases along with increasing porosity; increase of thickness can enhance the sound absorption coefficient at low frequencies but reduce it at high frequencies, but the contribution is not obvious for the whole sound absorption property, only that the migratory phenomena of maximal sound absorption coefficient towards low frequency appears.

Introduction

Aluminum foam is a kind of poroid late-model multifunctional material, which has fine characteristics such as low density, sound absorption, sound insulation, electromagnetic shielding, energy absorption, heat insulation and fireproofing, so the application field is quite wide. At present, aluminum foam under research can be divided into open-cell aluminum foam and closed-cell aluminum foam, and the former is prepared by seepage method, so cells are interconnected; the latter is prepared by foaming method, so cells are relatively independent. Closed-cell aluminum foam prepared by foaming method coats little, and is suitable for industrial production, so researchers pay more attention to closed-cell aluminum foam.

At present, researches abroad not only touch upon preparing technology, properties, but also include theory deducing, analog computation and so on [1-4], and this technology has been put to practical stage [5] in some countries. In China, researches on aluminum foam mainly focus on production processes [6,7] and different kinds of properties [8-12] recently, especially the static mechanical property [8,9]. Much progress has been made in recent years, and acoustic property of aluminum foam has also been reported, but most of which focus on acoustic property of open-cell aluminum foam prepared by seepage method [13-14], and there are few researches on acoustic property of closed-cell aluminum foam, even less the influential factors on sound absorption property. Therefore, in this paper sound absorption property of closed-cell aluminum foam was researched, sound absorption mechanism was discussed, and effects of porosity and thickness of closed-cell aluminum foam were also investigated.

Experimental

Preparation of Closed-cell Aluminum Foam.

The process has five step techniques [6]: (i) moltening alloy of aluminum-silicon and calcium (3%, the mass fraction) in furnace at 850□ (ii) adding titanium hydride (45 μ ms; 1.5%, the mass fraction) to the molten body at 680□ and stirring at a speed of 2000 r/min (iii) transferring the molten to the bubbly case (iv) pushing bubbly case to the maintaining furnace (650□) and foaming in it for 6 minutes (v) aluminum foam post processing. There are seven specimens prepared for sound absorption testing in this test, and parameters of which can be found in Table I.

Table I Parameters of closed-cell aluminum foam with different porosity and thickness

Species	Specimens	Density /g/cm³	Porosity /%	Thickness /mm	Main cell diameter /mm	Thickness of cell Wall /mm
Different porosity	1	0.85	67.3	20	3	0.6
	2	0.58	77.7	20	5	0.56
	3	0.51	80.4	20	6	0.44
	4	0.31	88.1	20	11	0.42
Different thickness	5	0.53	79.6	10	4	0.5
	6	0.53	79.6	20	4	0.5
	7	0.53	79.6	30	4	0.5

Typical specimen can be seen in Figure 1, which shows that the cell modality of specimens are uniform, wall of cells are joined by 'Y' type, and the angles are basically 120°.

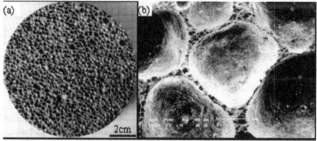

Figure1. : (a) Macroscopic and (b) microcosmic picture of specimen for sound absorption testing

Test of Sound Absorption

Sound absorption property of closed-cell aluminum foam was tested by standing wave tube at Institute of Acoustics, Chinese Academy of Sciences, and sound absorption coefficient of aluminum foam was also tested [15]. When sound wave meets the surface of closed-cell aluminum foam vertically, the traveling direction of reflected wave is against the direction of incidence wave, forming standing wave superposing one another. Normal incidence sound absorption coefficient can be figured out by measuring the maximal and minimal value of sound pressure. The measuring temperature and humidity are 24°C and 66% respectively; the range of frequency controlled is from 160Hz to 2000Hz. Figure 2 is the sketch map of measuring equipment.

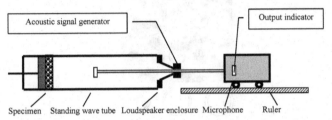

Figure 2. Sketch map of measuring sound absorption coefficient of closed-cell aluminum foam through standing wave tube

Results and Discussion

Sound Absorption Property of Aluminum Foam

The sound absorption property of closed-cell aluminum foam is shown in Figure 3, from which it can be seen that sound absorption property is much better under middle frequencies than those under low and high frequencies, sound absorption coefficient climbs when frequency increases from 160Hz to 800Hz and drops when frequency is added from 800Hz to 2000Hz, the highest sound absorption coefficient being 79%.

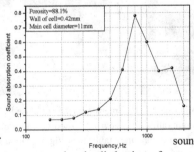

Figure 3. Curve of sound absorption characteristic of closed-cell aluminum foam

As the structure of aluminum foam is quite special, when the specimen is incised, different cell shapes appear on the surface of specimen, one of which is called Helmholtz resonator (see Figure 4 (b)) when the cell is incised only by a small hole [17]. As a result there are many Helmholtz resonators in parallel. While sound is imposed on the pipe orifices of Helmholtz resonators, the dimensions of pipe orifices being much smaller than that of the wavelength, each section of air in pipe orifices belongs to a small region within wavelength of λ. So the vibration behavior of them is considered uniform, which is to say that the air in pipe orifices as a whole vibrates just like a piston and frictionizes with the wall faces to consume sound energy.

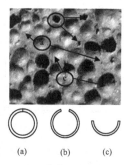

(a) (b) (c)

Figure 4. Surface pore modality of closed-cell aluminum foam

As for the air in cavity, when air column moves into the cavity, the air and the pressure in cavity will be compressed and increases because there is no way to go for the air, which causes the vibration of air and congregation of sound energy in cavity. The interior surface of aluminum foam is rough (see Figure 5(a)(b)), damping of itself large, the sound in which will be reflected and refracted many times owing to the resistance of flexural cell wall. That causes the friction between air and cell faces, therefore the cavity isn't suitable for store sound energy and the sound energy will transfer to heat energy to expand.

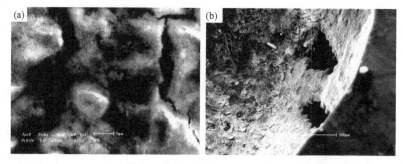

Figure 5. Crack (a) and micropore (b) of interior surface of closed-cell aluminum foam

In addition, owing to the high temperature of preparing closed-cell aluminum foam and cooling out of time, the cracks and micropores are inevitable (see Figure 5), so another main path to expand sound energy is through the cracks and micropores of aluminum foam. The damping characteristic of aluminum is not very good, but the loss coefficient value of aluminum foam is higher than aluminum by one order of magnitude at least [18]. The consumption of sound energy by aluminum foam is mainly through the friction between crack faces in the structure transmitting vibration energy to thermal energy and pushes in or pushes out air of micropore causing viscous consumption then dispersing into the surrounding environment.

Effect of Porosity

Figure 6 and Figure 7 are sound absorption properties of closed-cell aluminum foam with the same thickness but different porosity, from which it can be seen that the effect of porosity on sound absorption properties is obvious: sound absorption property of closed-cell aluminum foam with high porosity is better than that with low porosity. Results show that mean and highest sound absorption coefficients are 0.164 and 0.31 when porosity is 67.3%; the values reach 0.325 and 0.78 when porosity is 88.1%. Generally, increasing porosity can result in the decrease of Helmholtz resonator's number unit area and enhance of noise elimination capability of single Helmholtz resonator, so the effect of Helmholtz resonator on increasing sound absorption property owing to the increase of porosity can be neglected. The main reason why sound absorption efficiency increases with increasing porosity is that cells will be larger, wall of cell will be thinner, and opportunity of cracks and micropores will be larger with increasing porosity, so not only diffuse reflection of sound on the surface of closed-cell aluminum foam enhances and vibration of wall of cell resulting in interference noise elimination increase, but also frication and viscous expand resulted from cracks and micropores increase, so whole sound absorption property increases.

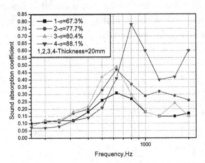

Figure 6. Curves of sound absorption characteristic of closed-cell aluminum foam of different porosity

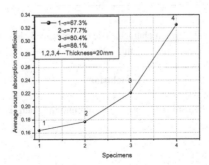

85

Figure 7. Average sound absorption coefficient of closed-cell aluminum foam of different
porosity

Effect of Thickness

Figure 8 is the characteristic curve of closed-cell aluminum foam with different thickness, which shows that sound absorption increases in low frequency but decreases in high frequency with increasing thickness of aluminum foam. The highest sound absorption coefficients for aluminum foam with thickness 10mm, 20mm and 30mm are 0.51, 0.52 and 0.49 respectively, and mean sound absorption coefficients are 0.22, 0.22 and 0.21, it follows that the effect of aluminum foam thickness on sound absorption property is not obvious, only the highest sound absorption coefficient transfers to low frequency, and test results coincide with the results gotten by Y. Wang [18].

Changes of sound absorption property of closed-cell aluminum foam affected by thickness is not obvious, the main reason is that if the wave impendence of air is not taken into consideration, the reflective coefficient and absorptive coefficient of material are fully controlled by surface impedance rate, while surface impedance rate of material is characterized directly by surface characteristic. Surface physical characteristics of materials chosen are alike basically, so the highest sound absorption coefficient and mean sound absorption coefficient do not change with the thickness added, as for the transference of highest sound absorption coefficient, the main reason is that sound absorption coefficient of laminae material in low frequency increases with increasing material thickness, so thicker material reaches the highest sound absorption coefficient firstly, and transference to low frequency of the highest sound absorption coefficient appears visually.

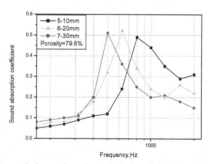

Figure 8. Curves of sound absorption characteristic of closed-cell aluminum foam of different
thickness

Conclusions

(1) Sound absorption mechanism of closed-cell aluminum foam is mainly controlled by

86

expendence of Helmholtz resonators, micropores, cracks, and vibration of wall of cell. Sound absorption tests of closed-cell aluminum foam with different porosity confirm that sound absorption coefficient of material increases with increasing porosity, and the highest and mean sound absorption coefficients are 0.78 and 0.325 respectively when porosity of material is 88.1%.

(2) Sound absorption tests of closed-cell aluminum foam with different thickness testify that sound absorption coefficient is high in low frequency, but low in high frequency, and the highest sound absorption coefficient transfers to low frequency with material's thickness added, but the highest and mean sound absorption coefficients do not change.

References

1. G.J. Davies and S. Zhen, "Metallic foams: their production, propertiesand applications," *J Mater Sci*, 18(1983), 1899-1911.
2. E.W. Andrews et al., "Size effects in ductile cellular solids. Part II: experimental results," *Int J Mech Sci*, 43(2001), 701-713.
3. Y. Sugimura et al., "On the mechanical performance of closed cell foams," *Acta Mater*, 45(1997), 5245-5259.
4. T. Anderson and E .Madenci, "Experimental investigation of low velocity impact characteristics of sandwich composites," *Compos Struct*, 50(2000), 239-247.
5. H.P. Degischer, and B. Kriszt, *Handbook of Cellular Metals: Production, Processing, Applications* (Austria: Wiley-VHH Press, 2002), 3.
6. G.C. YAO, et al., "The method of fabricating closed-cell foam aluminum by direct foaming in melt," *CN Patent*, (2001), 1320710.
7. B. Jiang et al., "Processing of open cell aluminum foams with tailored porous morphology," *Scripta Materialia*, 53(6)(2005), 781-785.
8. H.J. YU et al., "Influence of relative density on compressive behavior of Al-Si closed-cell aluminum foam," *J Northeastern University (Natural Science)*, 21(6)(2006), 1126-1129.
9. H.J. Yu, G.C. Yao and Liu Y H, "Research of tensile property of Al-Si closed-cell aluminum foam," *Trans Nonferrous Met Soc China, 2006*, 16(5)(in press).
10. N.Q. Zhao et al., "Effect of Y2OC on the mechanical properties of open cell aluminum foams," *Materials Letters*, 60(2006), 1665-1668.
11. H.J. Yu et al., "Electromagnetic shielding effectiveness of Al-Si closed-cell aluminum foam," *J Functional Mate*, 37(8)(2006), 1239-1241.
12. H.J. Yu et al., "Sound insulation property of Al-Si closed-cell aluminum foam sandwich panels," *Applied acoustics*, (2006)(in press).
13. L.C. Wang and F. Wang, "Research on sound absorption of aluminum foam," *Metallic Functional Materials*, 11(5)(2004),17-18.
14. H. Yu, L.G. Fang and Q.S. Yan, "The Fabrication of Porous Aluminum Alloys and the Testing of the Sound Absorbent Property," *Journal of Nanchang University (Engineering and Technology)*, 22(4)(2000), 10-14.
15. D.Y. Ma and H. Shen, *Acoustic handbook* (Beijing: Science press, 2004), 595.
16. D.Y. Ma, *Theoretic basis of modern acoustic* (Beijing: Science press, 2003), 237.

17. J. Kovacik, P. Busse and F. Simancik, *Metal Foams and Porous Metal Structures* (Bremen: MIT Verlag Press, 1999) 405.

18. Y. Wang, "Sound absorption characteristics of foamed aluminum," *Development and applicationg of materials*, 11(4)(1999), 15-18.

Aluminum Alloys
for Transportation, Packaging, Aerospace and Other Applications

Alloy
Development

Aluminum Alloys for Transportation, Packaging, Aeropsace, and Other Applications
Edited by Subodh K. Das, Weimin Yin
TMS (The Minerals, Metals & Materials Society), 2007

DEVELOPMENT OF SUPERPLASTICITY IN AN Al-Mg-Sc ALLOY

V. Jaya Prasad, S.S. Bhattacharya

Materials Testing Facility, Materials Forming Laboratory,
Department of Metallurgical & Materials Engineering, IIT Madras, Chennai, India – 600 036.

Keywords: Superplasticity, Al-Mg-Sc alloy, Thermomechanical treatments, Tensile deformation

Abstract

Superplasticity is the ability of polycrystalline materials to exhibit, in a relatively isotropic manner, large elongations prior to failure, under appropriate conditions of temperature and strain rate. This behaviour can be utilised in the shaping and forming of components, parts and structures that cannot be easily or economically produced from materials of normally limited ductility. Recent interest in replacing steel with aluminium in auto-body sheet-metal parts has led to Al-Mg based non-heat treatable alloys becoming potential candidates for automotive applications. In this work an Al-4wt.%Mg-0.25wt.%Sc alloy capable of exhibiting superplastic behaviour was developed. Using an appropriate casting schedule and a suitable thermomechanical processing route, fine-grained sheets of the alloy were produced. Tensile tests were carried out to characterize the high temperature deformation behaviour. It was demonstrated that the present alloy exhibits reasonably large elongations with a high sensitivity of the stress to the strain rate, typical of superplastic materials.

Introduction

Superplasticity is the process of deformation of materials that is capable of producing, under certain conditions, neck-free elongations of several hundreds of percent at very low loads [1]. In the case of conventional metals and alloys, superplastic behaviour is typically observed when the grain size is of the order of 1-20 μm, at temperatures above 0.5 T_m (where T_m is the absolute melting point) and at strain rates usually around 10^{-2} to 10^{-5} s^{-1}. One of the primary requirements for a superplastic material lies in its capacity to retain the stable ultra-fine grain size during isothermal deformation at high temperatures. Such materials show a high sensitivity of the stress to the strain rate. The strain rate sensitivity index, m, can be determined from isostructural, isothermal stress – strain rate plots for a material and is mathematically expressed by

$$m = \frac{\partial \ln \sigma}{\partial \ln \dot{\varepsilon}}\bigg|_{\varepsilon,d,T} \qquad (1)$$

where σ and $\dot{\varepsilon}$ are the true stress and true strain rate respectively at a constant value of strain (ε), grain size (d) and temperature (T). A high value of m confers resistance to localized (uncontrolled) deformation in a material. The incipient formation of a neck leads to a local increase of the strain rate, which leads to an increase of the flow stress in the necked region. If the strain-rate sensitivity is sufficiently high, the local flow stress increases to such an extent that further development of the neck is inhibited. While the value of m for a majority of superplastic materials lie in the range of 0.4 to 0.8, it is generally accepted that materials with m value greater than 0.3 can be superplastically deformed. The stress – strain rate relationship for superplastic flow is given by [2]

$$\sigma = K\dot{\varepsilon}^m \tag{2}$$

Besides high strain-rate sensitivity, a low rate of damage accumulation, such as cavitation, is required for the large plastic strains to be reached.

There has always been a strong interest in the use of superplastic deformation for the forming of components and the increase in commercial interest stems from the fact that superplastic materials exhibit low resistance to plastic flow as well as high plasticity. This combination of properties allows for the shaping and forming of complex shapes with a minimum expenditure of energy, while at the same time resulting in better service properties in the finished product due to uniform fine structure. Superplasticity can be used particularly effectively in applications requiring the forming of materials to large strain under tensile loading conditions and where a single superplastically formed component could replace complex welded assemblies. Superplastically formed parts find many applications, especially in the aerospace and surface transport industries. Superplastic forming has now been identified as a standard processing route in the aerospace industry. Aerospace applications of superplastic aluminium alloys are primarily in the fabrication of airframe control surfaces and small-scale structural elements where low weights and high stiffnesses are required. Non-aerospace applications of aluminium alloys include containers with complex surface profiles and decorative panels for internal and external cladding of buildings, window frames for trains and gate panels as well as for surface transport such as sports cars.

Many aluminium alloys, when thermomechanically processed into a fine equiaxed grain structure that remains relatively stable at the temperature of deformation, exhibit superplastic behaviour. Notable among these are the AA7475, AA6061 and AA2004 (Supral® series) alloys used extensively for structural applications. Currently, there is considerable interest in replacing steel with aluminium alloys in auto-body sheet-metal parts to improve fuel efficiency and environmental requirements. In addition, the requirements of low density, high strength, high toughness, good weldability and higher corrosion resistance have led to Al-Mg based non-heat treatable alloys becoming potential candidates for automotive applications.

To ensure an ultra-fine and stable grain size during superplastic deformation, alloying additions that result in a fine dispersion of submicrometer-sized second-phase particles are needed. These dispersoids pin the grain boundaries and avoid excessive cavity formation as well as do not coarsen at the deformation temperatures [3]. For example, the addition of Mn results in the formation of plate-like Al_6Mn-based particles that accomplish this role in case of the commonly used 5083 aluminum alloy [4-6]. Similarly, the addition of Sc or Zr in Al-Mg alloys result in small, coherent and spherical Al_3Sc or Al_3Zr particles, which are particularly effective in inhibiting grain growth at higher temperatures [7-12]. The addition of Sc to Al provides the highest increment of strengthening per atomic percent, compared to any other alloying element [13]. However, the absolute strength of Al-Sc alloys is relatively low as the maximum solubility of Sc in α-Al is only 0.38 wt.% [14]. Al-Sc alloys have a substantial age hardening response and age at a temperature higher than most conventional age hardenable Al alloys [13]. Alloying with Sc may inhibit or completely prevent recrystallisation in Al alloys [13,16] and so, cold or warm work prior to solution treatment may not result in any recrystallisation in such alloys [15]. Additional advantages of Sc addition in Al alloys result in enhanceed ductility, strength, toughness, fatigue life, corrosion resistance and the promotion of a more homogeneous distribution of solute elements within grains.

In this study, an attempt was made to develop a superplastic Al–Mg–Sc alloy through an appropriate casting schedule and thermomechanical treatment route. One of the goals in this exercise was to keep the casting and thermomechanical treatments procedures as simple as possible so that translation of the process to a commercial level becomes easily viable.

Experimental Programme

The composition for the Al-Mg-Sc alloy was selected in the range of 93-96 wt.% Al, 4-5 wt.% Mg and 0.25-0.2 wt.% Sc. The basic experimental procedure involved the casting of the alloy followed by a homogenization treatment, thermomechanical treatment in the form of cold and hot working, and tensile testing at high temperatures in order to characterize the superplastic deformation behaviour. Hardness measurements at each stage of the alloy preparation were made using a Vickers Hardness Testing unit with a pyramidal diamond indenter at a load of 2 kg. At least five readings were taken from different regions of the cast and processed alloy. At relevant stages microstructural studies using optical and electron microscopy were carried out. For optical metallography, the samples were prepared by following standard procedures of mechanical polishing on a series of emery papers followed by intial wet polishing with colloidal silica and finally with diamond paste. The polished specimens were etched with an etchant composed of 6ml HBF_4 + 26 ml HCl + 48 ml HNO_3 for about 30 seconds and washed thoroughly before observation under a metallurgical microscope.

Melting And Casting Of The Alloy

Commercially pure aluminium was used. Initial additions of the constituents were a little higher than the targeted nominal composition in order to offset oxidation and evaporation losses (which had been computed during trial attempts at melting). The addition of Sc was carried out in the form of a Al – 2 wt.% Sc master alloy, procured from a commercial source, whereas Mg addition was done in the elemental form. Appropriate amounts of the starting materials were melted in a clay–graphite crucible by heating to 993 K (720 °C) for 3 hours in an electric furnace using Kanthal heating elements. An argon atmosphere was maintained throughout the melting process. The molten metal was then poured (after degasification) into Direct Chill (DC) cast die moulds to produce ingots of dimensions approximately $200\times150\times25$ mm^3.

Homogenization Of The Cast Ingots

The cast alloy was homogenized at 753 K (480 °C) to reduce the degree of segregation and make the composition uniform. Initially, the direct chill cast stock was taken and heated to a temperature of 753 K (480 °C) for a period of 2 hrs. This process is also known in the trade as "preheating". After the time period of 2 hours, the ingot was cooled to the planned forging temperature of 723 K (450 °C) in the furnace and then taken out and allowed to cool in air. Following homogenization, the composition was checked and confirmed using an X-Ray Fluorescence Spectrometer (XRF).

Thermomechanical Treatment of The Alloy

The homogenized alloy ingot was hot forged at 723 K (450 °C) from a starting thickness of 20 mm to a final thickness of about 3.5 mm using a 200 kg Pneumatic Power hammer. The reduction achieved during this process was greater than about 80%. A 15% reduction was given to the hot forged ingot by using an experimental cold rolling mill in order to get a final thickness of 3 mm. Apart from hardness measurements, a room temperature tensile test was conducted on a specimen of (machined in accordance with ASTM E8 sub-size specifications) the alloy after thermomechanical treatment in order to determine the standard mechanical properties.

<u>High Temperature Tensile Testing</u>

High temperature tensile test specimens with their tensile axis parallel to the rolling direction and with gauge dimensions of 12.5 mm length and 4 mm width were used for the high temperature tensile tests at temperatures ranging from 748–823 K (475–550 °C) at 25° intervals. A schematic of the tensile test specimen is shown in Figure 1. Prior to testing the specimens were mechanically polished to remove fine scratches from the specimen surface, particularly in the gauge portion.

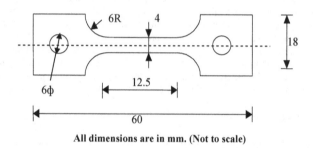

All dimensions are in mm. (Not to scale)

Fig.1. High temperature tensile specimen (schematic)

For the high temperature uniaxial tensile testing campaign, a microprocessor controlled screw-driven electromechanical testing machine (make: Schenck-Trebel) of 250 kN capacity was used. The machine is direct current driven and fully computerized for conducting experiments under displacement, force and strain controlled modes as well as for data acquisition. The machine is also equipped with a thyristor controlled two zone split furnace, which ensures a uniform temperature zone of 250 mm with an accuracy of $\pm$ 3°. In this study a 1 kN load cell was used for measuring the loads. Each sample was held at the testing temperature for about 10 minutes in order to reach thermal equilibrium.

The tensile tests were conducted at different constant Cross Head Speeds (CHS), during which the true strain rate continued to decrease as the specimen elongated. At each temperature, five different initial strain rates, viz., 1.04×10^{-4} s^{-1}, 5.0×0^{-4} s^{-1}, 2.0×10^{-3} s^{-1}, and 5.0×10^{-3} s^{-1} (corresponding to CHS of 0.075 mm min^{-1}, 0.375 mm min^{-1}, 1.5 mm min^{-1}, and 3.0 mm min^{-1} respectively) were employed. All the tests were carried out until fracture and the load-elongation data acquired. The true stress (σ_t) and true strain (ε_t) values were calculated from the load-elongation data. The fracture surfaces were observed using a Scanning Electron Microscope (SEM).

Results And Discussions

The typical as-cast structure was lost after homogenization and the microstructure is shown in Figure 3. The two dimensional grain size was found to be in the range of 39 μm. Table I gives the chemical composition of the alloy. The presence of Mn, Si and Fe was attributed to the impurities present in commercially pure aluminium. It is, however, noteworthy that the target composition was more or less achieved.

Table I: Chemical composition (in weight %) of the alloy

Element	Al	Mg	Sc	Mn	Si	Ti	Fe	Na
Al Alloy	93.25	4.17	0.23	0.34	0.89	0.02	0.37	0.09

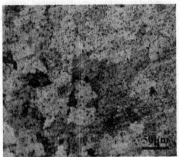

Figure 2: Microstructure after homogenization

The microstructure after thermomechanical treatment is shown in Figure 4. The two dimensional grain size (obtained by the linear intercept method) worked out to about 11 μm with slightly elongated grains having an aspect ratio of 1.4.

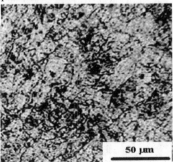

Figure 3: Microstructure of the alloy after thermomechanical processing

The room temperature mechanical test yielded an Ultimate Tensile Strength of 410 MPa with an elongation to failure of 22%. An average Vicker's Hardness Number (at 2 kg load) of 97 was obtained. The results of the high temperature tensile tests in terms of the maximum stress and the percentage elongation to failure are presented in Table II. From the data it is seen that the alloy exhibited reasonably good tensile elongations in most of the cases. In all the cases, a nearly steady-state condition was observed, although some work hardening was evident. Strain softening following the stress maximum occurred until failure and nearly uniform deformation took place within the gauge length. No visible necking took place around the fracture indicating that the deformation was homogeneous and that necking was restrained. [17,18]. Increase in temperature led to a shift of the peak stress and a reduction in the initial work hardening. At higher deformation rates the work hardening rate saturated early during deformation, whereas the slower deformation rates showed a more sustained hardening rate maintained to higher strain levels. In general, the alloy showed an increase in strain to failure with a decrease in initial strain rate with the slope of the stress – strain curve increasing with decreasing initial strain rate, delaying the attainment of maximum stress. At high strain rates, the hardening component was attributed essentially due to dislocation processes such as pile-ups, tangles and sub-grain formations, which at higher temperatures saturated quickly [19]. On the other hand, at the lower strain rate of 1×10^{-4} s^{-1} the steady deformation was attributed primarily to the optimal and near-optimal superplastic flow regime. The true stress – true strain plot at 823 K and different initial strain rates is depicted in Figure 4.

Table II: Results of the high temperature tensile tests

T (K)	Strain rate (s^{-1})	Max. stress (MPa)	Elongation (%)
823	1.0×10^{-4}	3.5	212
	5.0×10^{-4}	4.2	185
	2.0×10^{-3}	7.1	170
	5.0×10^{-3}	11.2	146
798	1.0×10^{-4}	4.1	186
	5.0×10^{-4}	6.3	166
	2.0×10^{-3}	12.2	136
	5.0×10^{-3}	14.9	113
773	1.0×10^{-4}	5.6	162
	5.0×10^{-4}	10.1	150
	2.0×10^{-3}	13.9	120
	5.0×10^{-3}	18.7	91
748	1.0×10^{-4}	8.6	153
	5.0×10^{-4}	8.8	132
	2.0×10^{-3}	14.9	104
	5.0×10^{-3}	29.8	88

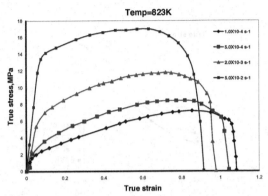

Figure 4: True stress – true strain plot at 823 K at different initial strain rates

True stress – true strain rate plots at constant true strain levels were generated at different true strain levels. A sigmoidal relationship between the flow stress and strain rate was observed for all the cases and the three well-known regions of superplastic flow could be identified. The variation of the flow stress (at a true strain of 0.4) as a function of the instantaneous strain rate, plotted on a double logarithmic scale, is shown in Figure 5. The slope of the ln σ - ln ε̇ curve gave the instantaneous value of the strain rate sensitivity index, m (as given by Equation 1). An average value of the strain rate sensitivity index was also obtained by assuming a linear fit between ln σ - ln ε̇ at a given temperatures and strain level. In most of the cases, the value of m was more than 0.3 indicating the superplastic tendency of the alloy. A maximum average m value of 0.41 was obtained at 823 K.

Microstructural studies of the tested specimens were carried out at the gauge and grip portions in order to distinguish between dynamic (gauge) and static (grip) effects. Figure 6 gives the microstructure of a specimen tested at 823 K and 5.0×10^{-4} s^{-1}. The average two dimensional grain size at the gauge portion was 21 μm and at the grip portion was 19 μm. The extent of grain growth in all the other cases was considerably lower.

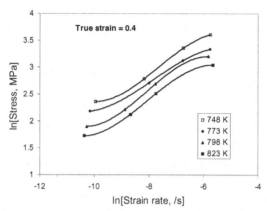

Figure 5: Variation of flow stress with strain rate at a true strain level of 0.4

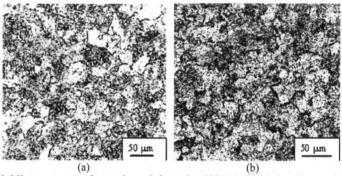

(a)   (b)

Figure 6: Microstructures of a specimen deformed at 823 K and initial strain rate of 5.0×10^{-4} s^{-1} at the (a) gauge portion and (b) grip portion.

When a superplastic material fails during tensile deformation it is either the result of unstable plastic flow or a consequence of the growth and interlinkage of internally nucleated voids [20]. At higher temperatures, grain boundary cavity formation caused by grain sliding takes place and results in lower ductility values [12]. Figure 7 depicts the fracture surface (SEM image) in case of a specimen tested at 748 K at an initial strain rate of 1.0×10^{-4} s^{-1} and Figure 8 depicts the fracture surface of a specimen tested at 823 K and initial strain rate of 2.0×10^{-3} s^{-1}. A large number of big and small dimples and several tear ridges were observed. The dimple sizes increased with temperature while the width of the tear ridges decreased. Some individual grains are also seen inside the cavities and also along the walls of the cavities. At higher temperatures, large cavities interlinked by long tear ligaments were observed. Failure of some of these tear ligaments led to coalescence of cavities. This indicated that fracture took place at this temperature by cavity nucleation and growth. Thus it could be concluded that the specimen had basically fractured intergranularly, the fracture manner being a combination of cleavage or quasi-cleavage between grain boundaries and tearing fracture modes.

Figure 7: Fracture surface of a specimen tested at 748 K and an initial strain rate of 1×10^{-4} s^{-1}.

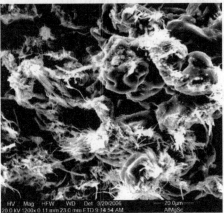

Figure 8: Fracture surface of a specimen tested at 823 K and an initial strain rate of 2×10^{-3} s^{-1}.

Concluding Remarks

In this study, an effort was made towards the development of a superplastic Al–4wt.%Mg–0.25wt.%Sc alloy. An appropriate casting schedule for the preparation of the alloy was established, which resulted in a composition reasonably close to the nominal value. A suitable thermomechanical treatment route consisting of 80% reduction by hot forging and 15% reduction by cold rolling resulted in a 3 mm thick strip of the alloy with an average and nearly equiaxed grain size of 11μm. High temperature tensile tests in the temperature range of 748 – 823 K and strain rate range of $1 \times 10^{-4} - 5 \times 10^{-3}$ s^{-1} resulted in elongations of more than about 130 % under most testing conditions. In addition, the strain rate sensitivity index, m, was more than 0.3 under most conditions, which clearly indicated the superplastic nature of the alloy. A maximum elongation of 212 % at 823 K and 1.0×10^{-4} s^{-1} was observed. The corresponding values of the strain rate sensitivity index under these deformation conditions were the highest. The optimal superplastic temperature and strain rate ranges were found to be 798-823 K and 5.0×10^{-4} to 1.04×10^{-4} s^{-1}. Fractographic studies showed evidence for grain boundary sliding which is the predominant mechanism during superplastic flow. Intergranular fracture as a combination of quasi-cleavage between grain boundaries and tearing fracture modes were observed.

Acknowledgement

The authors would like to thank the Office of Naval Research, USA (Award No. N00014-97-1-0993) for financial support under an Indo-US Project.

References

1. K.A. Padmanabhan and G.J. Davies, Superplasticity, Springer-Verlag, Berlin, 1980.
2. W.A. Backofen, I.R. Turner and D.H. Avery, Trans. Am. Soc. Metals, 57 (1964) 980.
3. J.S. Vetrano, S.M. Bruemmer, L.M. Pawlowski and I.M. Robertson, Mater. Sci. Eng. A, 238 (1997), 101.
4. F. Li, W.T. Roberts and P.S. Bate, Acta Mater., 44 (1996), 217.
5. D.H. Bae and A.K. Ghosh, Acta Mater., 48 (2000), 1207.
6. E. Usui, T. Inaba, N. Shinano, Z. Metallk., 77 (1986), 179.
7. R.R. Sawtell and C.L. Jensen, Metall. Trans., 21A (1990), 421.
8. T.G. Nieh, L.M. Hsiung, J. Wadsworth and R. Kaibyshev, Acta Mater., 46 (1998), 2789.
9. F. Musin, R. Kaibyshev, Y. Motohashi and G. Itoh, Scripta Mater., 50 (2004), 511.
10. S.J. Hales and T.R. McNelley, Acta Metall., 36 (1988), 1229.
11. C.A. Lavender, J.S. Vetrano, M.T. Smith, S.M. Bruemmer and C.H. Hamilton, Scripta Metall. Mater., 30 (1994), 565.
12. R. Kaibyshev, F. Musin, D.R. Lesuer and T.G. Nieh, Mater. Sci. Eng., A 342 (2003), 169.
13. L.S. Kramer and W.T. Tack, Adv. Mater. Proc., 10 (1997) 23.
14. J.L. Murray, J. Phase Equil., 19 (1998) 380.
15. B. Irving, Weld. J., 76 (7) (1997) 53.
16. V. Ocenasek and M. Slamova, Mater. Charact., 47 (2001) 157.
17. J.W. Edington, K.N. Melton and C.P. Cutler, Progr. Mater. Sci., 21 (1976) 61.
18. Y.H. Wei, Q.D. Wang, Y.P. Zhu, H.T. Zhou, W.J. Ding, Y. Chino and M. Mabuchi, Mater. Sci. Eng., A360 (1-2) (2003) 107.
19. R. Verma , A.K. Ghosh, S. Kim and C. Kim, Mater. Sci. Eng., A191 (1-2) (1995) 143.
20. D. Ravi Kumar and K. Swaminathan, Mater. High Temp., 16 (4) (1999) 161.

PRODUCTION OF AA6061 BILLETS FOR THIXOFORGING

Yucel Birol

Materials Institute, Marmara Research Center, TUBITAK, Kocaeli, Turkey

Keywords: Semi-solid forming; Thermomechanical processing; Aluminium alloys.

Abstract

The extruded AA6061 alloy recrys tallized during heating to the sem isolid state, producing nearly equiaxed polyg onal g rains with h igh angle boun daries. Meltin g started at triple junctions during subsequent isotherm al heating and progressed along grain boundaries. The low m elting point eutectic phas e an d the hi gh angle grain boundaries introduced through recrystallization both p romoted the grain bo undary m elting proces s. Com plete grain boundary wetting, a perfectly cylin drical slug shape and an av erage grain size below 100 microns were obtained after 5 minutes at 640°C for an extrusion ratio of 14:1.

Introduction

Near-net shaping of alum inium components in the sem isolid state is already a commercial manufacturing route which produ ces m illions of autom otive par ts a nnually [1]. The key feature tha t perm its the sem isolid shaping of alloys is a dendrite-fr ee m icrostructure, with globular α-Al grains suspended in a liquid m atrix, which m ay be handled like solids, but flows readily when sheared. Such slurries m ay be shaped via casting or forging to produce components with superior microstructural features and mechanical properties [2-4].

Thixoformed com ponents are alm ost always made from the A356 and A357 casting alloys which provide good fluidity and castability ow ing to a relatively high volum e of Al-Si eutectic. Thixoforming wrought alum inium alloys, which offer better m echanical properties than the cast grades, on the other hand, suffer se veral drawbacks [5]. Much lower volum e of eutectic phase, narrow freezing range and high sensitivity of the liquid fraction to temperature fluctuations are typical in these alloys, and leave a processing window which may be as small as several degrees [6]. Nevertheless, there is significant interest in thixoform ing wrought aluminium alloys.

Among several m ethods frequently used to pr oduce wrought alum inium alloy feedstock for thixoforming [6-13], the therm omechanical proce ssing route is particular ly attractive owing to the commercial availa bility of aluminum billets with di fferent compositions and e xtrusion ratios. This process involves the partial m elting of a hea vily deformed alloy in order to obtain a fine equiaxed m icrostructure and is well suited to sm all diameter m aterials [9,11]. The transform ation of t he deform ed dendritic m icrostructure into an equiaxed one occurs through recovery and recrystallization during heating [14]. New, strain-free grains with high-angle boundaries thus for med, are penetrated by the liquid phase once the temperature increases above the solidus tem perature. In the present work, extrude d bars of the AA6061 alloy were isothermally held in the semisolid state and subsequently quenched. The effect of reheating treatm ents on the evolution of m icro and m acro stru ctural featu res were investigated to identify the optimum practice.

Table 1. Chemical composition of the A6061 alloy used in the present work.

Si	Fe	Cu	Mn	Mg	Ti	Cr	Al
0.666	0.183	0.186	0.026	0.709	0.010	0.062	98.14

Experimental Procedures

AA6061 alloy (chem ical com position given in Table 1) used in the present work was cast industrially with a state-of-the art hot top air-sl ip vertical billet caster in the form of 7400mm long, 208mm dia meter billet. The as-received billet was first reduced in diam eter for the laboratory press by rem elting a nd c asting into a cylindrical permanent mold. The 100mm diameter billet thus obtained was heated to 460 °C and was subsequently extruded into 27mm diameter bars (extrusion ratio 14:1) in a hydra ulic press. 40mm l ong slugs were sectioned from the extruded bars and a m edium frequency induction coil (9.6 kHz, 50 kW) was used to heat thes e slugs into the sem isolid range. Temperature was m onitored with a K-type thermocouple inserted in a 3mm di ameter hole drilled at the center of the slugs. Measures were taken to achieve rapid heating to the se misolid range to assur e s mall recrystallized grains. The average h eating rate was approxim ately 150 °C/min. The slugs w ere then isothermally held in this temperature range for upto 30 m inutes and then quenched in water. The evolution of microstructure during holding in the semisolid state was investigated.

The heat flow vs. te mperature curves obtained by Differential Scanning Calorimetry (DSC) were used to calcu late the change in liquid -solid fractions with tem perature. Tem peratures for reheatin g experim ents were th en estim ated from the latter. The reheated slugs were sectioned and prepared with standard metallographic practices and finally etched with a %0,5 HF solution before they were examined with an optical microscope. Dark field im aging was used to better elucidate the ch anges in microstructural features including the very fine Mg_2Si precipitates. While the entire section of the reheated slugs was analyzed, the micro structural features were reported only for a 25mm [2] area which sits halfway between the center and the edge of the section.

Results and Discussion

α-Al dendrites and the inte rmetallic particle s lo cated a t in terdendritic sites in th e star ting AA6061 billet are read ily iden tified in Fig. 1a. The m icrostructure of the extrud ed bar is typical of hot-worked m etals with inte rmetallic particles a ligned in th e extru sion direc tion (Fig. 1b). The extruded m icrostructure is do minated by fibrous grains and the Fe-based intermetallic particles h ave undergo ne substantial fragm entation. Air-quenching from the press has produced a considerable a mount of Mg $_2$Si precipitation. Exam ination of the transverse and longitudinal se ctions showed no evidence of recrystallization. This was confirmed by hardness measurements which showed the extruded bar to be approximately 15 HV harder than the as-cast bill et, implying that the deform ation introduced during extrusion was largely retained. This deform ation was f ound to suffi ce to fully recrystallize the grain structure in the extruded bars during reheating. A separate cold work, often used in the Strain Induced Melt Activated (SIMA) route subsequent to the hot deformation, was thus omitted in the present work.

The isotherm al heating tem perature for the ex truded bars in the sem isolid range to achieve partial rem elting was es timated from the solid f raction vs. tem perature curve (Fig. 2). It is generally claim ed that 30 to 50% liquid is ne eded in the feedstock for thixoform ing [15], indicating a thixoform ing tem perature range of approxim ately 622-641 °C for the presen t alloy. W hile the change in solid fraction with tem perature (dFs/dT) is m uch s maller near

102

630°C (only as m uch as 0.004 °C⁻¹), reheating temperatures, hi gh enough to m elt, in addition to the eutectic

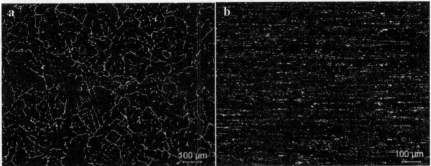

Figure 1. The dark field optic al m icrographs of (a) the as-cast and (b) extruded AA6061 billet.

phase, part of the pri mary α-Al phase, m ay be needed. Tem peratures between 630-640 °C were thus judged to be the most appropriate. Given enough time, isothermal heating at 640°C is expected to provide as much as 48% liquid phase.

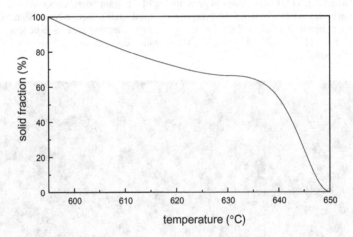

Figure 2. The change in solid fraction with temperature for the AA6061 alloy.

The extruded 6061 bars have recr ystallized during heating to the semisolid range, producing nearly equiaxed polygonal grains with high angle boundaries. Several additional changes were noted during sub sequent isoth ermal heat ing. Meltin g started at triple junctions and progressed along grain boundaries (Fig. 3). The low m elting point eutectic phase and the high angle grain boundaries introduced by recrystallization both promoted the grain boundary melting process. It is inferred from the m icrographs that the intragranular regions have been gradually vacated while the decoration of th e grain boundaries with the liqu id phase has advanced to produce a nearly co mplete in tergranular network with increasing heating temperature and holding tim e (Fig. 4). These microstructural changes can be acco unted for by the solutionizing of the second phase particles (AlFeSi in termetallics and Mg ₂Si precipitates) insid e the gra ins and the dif fusion of solute Fe, Mg and Si to the grain boundaries, thus increasing the fr action of eutectic phases whic h have m elted and penetrated between the grains.

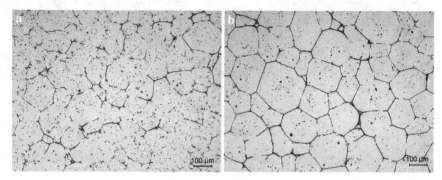

Figure 3. T he microstructures of extruded bar rapidly heated to 630 °C and isthe rmally held at this temperature for (a) 5 and (b) 25 minutes.

While holding at 630 °C for upto 15 m inutes has failed to produce enough liquid, holding for 10 m inutes at 635°C barely provided the liq uid phase to penetrate between the grain boundaries (Fig. 4). However, complete wettin g of the grain boundaries was evid ent after only 5 m inutes at 640°C. W ith the exception of a few grains, there were no signs of liquid entrapment inside the grains. The liquid content after reheating appears to be less th an that estimated from the liquid- solid fraction vs temperature curve in Fig. 2. This is believed to be due to the solidification of

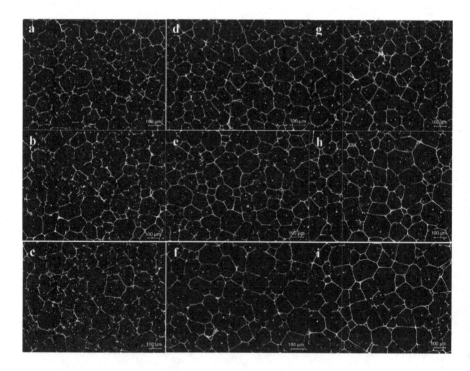

Figure 4. The m icrostructures of the extruded ba rs af ter they were held iso thermally at 630°C for (a) 5, (b) 10 and (c) 15 m inutes; at 635 °C for (d) 5, (e) 10 and (f) 15 m inutes; at 640°C for (g) 5, (h) 10 and (I) 15 minutes.

liquid on the α-Al grains [16] and to the liquid segregation across the cross section of the slug during subsequent quenching [5].

It is clear from the foregoing that both the temperature and time of the reheating process need to be fine tuned to have all the grain boundari es wet by the liquid phase. Such optim ization must take into account not only the liquid fraction and the grain boundary wetting but also the structural coarsening and the slug shape. W hile the grain boundary wetting im proves with increasing reheating temperatures (Fig. 4), there is a lim it for reheating tem perature, to avoid the distortion of the slug for trouble-free handling. Likewise, t oo long an isotherm al heating will cause s tructural coarsening and is not attrac tive cost-wise either. A short hold ing time, on the other hand, leads to incom plete globuriza tion of the grains im pairing the thixotropic properties and die filling ability [4].

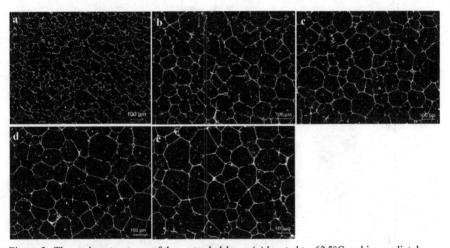

Figure 5. The m icrostructures of th e extruded bars, (a) h eated to 63 5°C and im mediately quenched in water; held isothermally at 635°C for (b) 5, (c) 10, (d) 15 and (e) 20 minutes and then quenched in water.

Fortunately, the coarsening of α-Al grains, driven by the grain boundary energy, was found not to be a major risk in the presen t case. The rate of coarsening was prominent only during the first 5 minutes of isotherm al heating when the grain size almost doubled, as evidenced by a series of micrographs, which show the evol ution of microstructure with time at 635°C (Fig. 5). The grain coarsening is shown in term s of $(D^2-D_0^2)$ in Fig. 8a where D and D $_0$ are the grain sizes at time t and at the s tart of isothermal heating (t=0). The average grain sizes were invariably below 100 microns for isotherm al h eating treatm ents that lasted 10 m inutes or shorter. Considering th at 10 m inutes is the relevant rehe ating duration in the thixo forming route and th at an averag e grain s ize less than 100 m icrons is desired for the thixofor ming operation, the grain sizes in the present case are all acceptable. The average shape factors of the grains were nearly constant in the entire holding range, suggesting that the globurization did not im prove at all after 5 m in in the sem isolid state (F ig. 6b). This is due to the f act that

105

the majority of the grain s were already equiax ed at the start of isotherm al holding owing to the recrystallization process that occurred during heating.

The load sustain ing ability of the slugs was gradually degraded with increasing reheating temperatures and hold ing tim es, eventually leadi ng to an "eleph ant f oot" shape, typical of over-

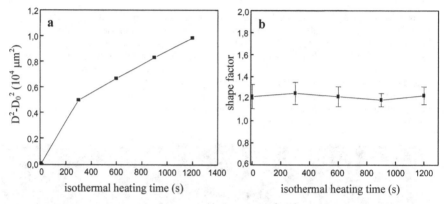

Figure 6. The grain coarsening, (a) expressed in term s of $D^2-D_0^2$ and (b) the change in shape factor of the grains with isothermal heating time.

Figure 7. The shape of the slugs sectioned from   extruded bars after they were held isothermally at 630 °C for (a) 5, (b) 10 and (c) 15 m inutes; at 635 °C for (d) 5, (e) 10 and (f) 15 minutes; at 640 °C for (g) 5, (h) 10 and (I) 15 minutes.

heated material. Slugs held longer than 20 min at 630°C and longer than 5 min at 635-640°C were deformed under their own weight (Fig. 7).

The critical properties o f the feedstock m aterial discussed so far are summ arized in Table 2 for each set of isothermal heating temperature and time. The first of the two signs in each cell indicates the extent of grain boundary wetting. "+" stands for complete wetting while "–"implies any state less than complete. The second sign is for the slug shape after reheating. Here again, "+" stan ds for a pe rfectly cylindrical shape whereas even a slight distortion is assigned "–". The average grain size for each isothermal heating treatment is also included in Table 2. Complete grain boundary w etting, a perfectly cylindrical slug shape and an average grain size less than 100 m icrons are obtained wh en the extruded bar is held at 640°C for 5 minutes following extrusion. Sim   ilar range s were reported for w   rought 6X XX alloys [5,8,17]. While the narrow pro   cessing window was reported   to cause difficulties in establishing a hom   ogeneous solid–liquid di   stribution in the slugs for conventional thixocasting [17], 6061, 6063, 6082 and 6106 alloys   were found to be thixoformable even when the starting structure was not perfectly   thixotropic [18,19]. The nearly globular microstructures obtained by therm o m echanical processing in the present work (Fig. 3) are thus expected to be highly formable in a thixoforging operation.

Table 2. The state of grain boundary   wetting (com plete:+; incom plete: –), slu g shape (cylindrical:+; deformed: –) and average grain size of extruded bars after isotherm al heating at various temperatures for various times.

Isothermal heating time (min)	isothermal heating temperature (°C)		
	630	635	640
5	–/+/78	–/+/89	+/+/95
10	–/+/94	+/–/98	+/–/104
15	–/+/100	+/–/106	+/–/109

Conclusions

The extruded 6061 alloy recrystallized during heating to the semisolid state, producing nearly equiaxed polygonal grains with high angle bounda ries. Melting started during isotherm al heating at tr iple junc tions and prog ressed along grain boundaries. The low m elting point eutectic phase and the high angle grain bound   aries introduced by re crystallization both promoted the grain boundary melting process. Complete wetting of the grain boundaries was evident after only 5 m inutes at 640°C. The average g rain sizes were invariably below 100 microns for all isotherm al heating treatm ents that lasted 10 m   inutes or less. Longer isothermal heating is claim ed to be unnecessary since the globurization of the grains did not improve aft er 5 m in. The desired feedstock   characteristics, com prising com plete gra in boundary wetting, a perfectly cylindrical slug sh ape and an average grain size less than 100 microns were obtained after 5 minutes at 640°C.

Acknowledgements

F. Alageyik and O. Cak ır are thanked for their help in th e experiments. The AA6061 billet was supplied by ASAS Alum inium Co. Thi s work was funded by the State Planning Organization of Turkey.

REFERENCES

1.   M. Garat, S. Blais, C. Pluchon, W .R. Loue. in: A.K. Bhasin, J.J. Moore, K.P. Young, S. Midson (Eds.), Proceedings of the Fift h International Conference on Se mi-Solid Processing of Alloys and Composites. Golden, Colorado, (1998) 17.
2.   M. C. Flemings, Metallurgical Transactions A, 22A (1991) 957.
3.   D. H. Kirkwood, International Materials Reviews, 39 (1994) 173.
4.   Z. Fan, International Material Reviews, 47 (2002) 1.
5.   D. Liu, H.V. Atkinson, P. Ka pranos, W . Jirattiticharoean and H. Jones, Materials Science and Engineering A61 (2003) 213.
6.   X. D. Zhang, T. A. Chadwick, J. D. Bryant, Materials Science Forum, 331-337 (2000) 247.
7.   S.Y. Lee and S. Oh, Journal of Ma terials Processing Technology, 130-131 (2002) 587.
8.   N. Saito, T. Naka mura, T. Ohtani, M. Kuroki, E. Masuda, T.Idegomori, JSAE Review, 22 (2001) 29.
9.   K.P. Young, C.P. Kyonka and J.A. Courtois, US Patent 4,415,374, 1983.
10. T.   Haga and P. Kapranos, Journal of Materials Processing Technology, 130–131 (2002) 594.
11.   M.P. Kenney, J.A. Courtois, R.D. E vans, G.M. Farrior, C.P. Kyonka, A.A. Koch and K.P. Young, "Sem isolid Metal Casting and Forging", *Metals Handbook*, Ninth Edition, Vol.15, (1998) 327.
12.   E. Tzimas and A. Zavaliangos, Materials Science and Engineering, A289 (2000) 217.
13.   Y. Birol, Materials Science Forum, 519-521 (2006) 1919.
14.   R.D. Doherty, H. Lee, E. Feest, Materials Scence and Eninering 65 (1984) 181.
15.   H.V. Atkinson, P. Kapranos, D. Liu, S. A. Chayong and D.H. Kirkwood, Materials Science Forum, 396-402 (2002) 131.
16.   E. Tzimas and A. Zavaliangos, Materials Science and Engineering, A289 (2000) 228.
17.   H. Kaufmann, H. W abusseg, P.J. Uggowitzer, in: G.L. Chiarm etta, M. Rosso (Eds.), Proceedings of the Sixth International Conference on Semi-Solid Processing of Alloys and Com posites, Turin, Italy, 27–29 Septem ber 2000, Edim et Spa, Brescia, Italy, (2000) 457.
18.   G. Tausig, K. Xia, in: A.K. Bhas in, J.J. Moore, K.P. Young, S. Midson (Eds.), Proceedings of the Fi fth International Conference on Semi-Solid Processing of Alloys and Com posites, Golden, CO, USA, 23–25 June 1998, Colorado S chool of Mines, Golden, CO, USA, (1998) 473.
19.   G. Tausig, in: G.L. Chiarm etta, M. Rosso (Eds.), Proceedings of the Sixth International Conference on Se mi-Solid Processing of Alloys and Composites, Turin, Italy, 27–29 September 2000, Edimet Spa, Brescia, Italy, (2000) 489.

Aluminum Alloys for Transportation, Packaging, Aeropsace, and Other Applications
Edited by Subodh K. Das, Weimin Yin
TMS (The Minerals, Metals & Materials Society), 2007

Application of a New Constitutive Model for the FE Simulation of Local Hot Forming of Age Hardening Aluminium Alloys

S. Gouttebroze[1], A. Mo[1], and H.G. Fjær[2]

[1]SINTEF, Material and Chemistry, N-0314 Oslo, Norway
[2]Institute for Energy Technology, N-2027 Kjeller, Norway

Keywords: hot forming, internal variable constitutive model, precipitate hardening, dislocation hardening/recovery, aluminium alloy

Abstract

In local hot forming processes, the material is subjected to local heating while being simultaneously deformed. The softening behaviour that age hardening alloys exhibit at elevated temperatures can then be exploited in order to effectively carry out certain critical forming operations. A new constitutive model is presented that relates the flow stress to the strain rate, temperature, and microstructure in age hardening aluminium alloys. The model accounts for the accumulation and annihilation of dislocations, as well as for the changes in the volume fraction and size distribution of the hardening precipitates. The model is applied in case studies to which histories of strain rates and temperatures typical for local hot forming are input. The calculated flow stress history is compared to similar histories obtained by a simpler, and in FE codes more commonly used, material behaviour modelling. Consequences of not accounting directly for the mentioned microstructural aspects are discussed.

Introduction

In the local hot forming process, the material is subjected to local heating while being simultaneously deformed, either thermally induced or by an additional mechanical load. The softening behaviour that age hardening alloys exhibit at elevated temperatures can then be exploited in order to effectively carry out certain critical forming operations and/or controlling the local microstructure [1,2].

The advanced use of mathematical modelling and associated finite element (FE) codes can be a great help when the benefits of local hot forming is to be exploited. A critical part of such models, though, are the constitutive equations by which the flow stress associated with the viscoplastic deformation is related to the viscoplastic strain rate, temperature, and microstructure.

The challenge associated with establishing accurate and reliable constitutive equations for modelling hot forming of age hardening aluminium alloys is that the flow stress changes not only due to the creation and annihilation of dislocations (i.e., strain hardening and recovery [3,4]), but also due to changes in the volume fraction and size distribution of the hardening precipitates [5]. Both factors have a significant influence on the value of the local flow stress due to complex interactions between the precipitates and the dislocations [6]. Recently, the authors presented a new constitutive model for the application in FE simulations of local hot forming of aluminium 6xxx alloys [7]. By means of internal variables, this model accounts at the continuum level for the previously mentioned effects that the evolving microstructure may have upon the viscoplastic flow stress.

The present study compares the new internal variable model with the more traditional approach based upon using so-called hardening curves in the FE modelling of metal forming operations (see for example Reference 8). By the latter methodology, a set of empirical stress-strain curves are established by e.g. tensile testing of the actual alloy at relevant temperatures, strain rates and levels of strain, and in the subsequent FE simulation, the flow stress during viscoplastic deformation is calculated by interpolating between the stress-strain curves. However, although each mechanical test may quite accurately quantify the viscoplastic flow stress for the conditions applying in the test, the methodology does not necessarily account very accurately for how the microstructure evolution in a real forming operation with varying temperature and strain rate affects the flow stress.

The new constitutive model and its empirical basis for parameter identification are briefly presented in the next section. A set of stress-strain material curves is furthermore generated by means of the model. Three typical evolutions of temperature and strain rates in material points of a work piece undergoing local hot forming are then used as input when calculating the flow stress during viscoplastic deformation by means of the new constitutive model and the hardening curve approach, respectively; the latter using the stress-strain curves generated by the new constitutive model. The flow stresses resulting from the two approaches are then compared and discussed.

Constitutive models

The new constitutive model

The new model combines existing models accounting for strain hardening and recovery [4,6] and the precipitation of particles from solute elements [5,9–11] and their contribution to the yield stress [12,13], and can thus be divided into different modules as indicated in Figure 1. With all symbols defined in Table 2, the corresponding equations are presented in Table 1, and a brief summary is given in what follows. Reference 7 should be consulted for the details.

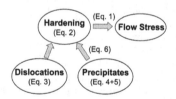

Figure 1: Modular structure of the constitutive model

The viscoplastic equivalent strain rate and equivalent flow stress are related to each other by a kinetic equation (Equation 1) in which both forest dislocations and precipitates contribute to the hardening. Please note that the parameter m is strongly dependent on temperature, while $\dot{\epsilon}_0$ is assumed constant, and that the parameter χ is introduced in order to keep the dislocation density reasonably low at high temperatures. Two different mechanisms contribute to the hardening (Equation 2), namely the development of forest dislocations by dislocation-dislocation interactions, and the solute elements and precipitates. The associated stress contributions are in the proceeding referred to as dislocation stress ($\sigma_\perp$) proportional to the square root of the dislocation density, and precipitate stress (σ_{ppt}), respectively.

The first two terms in the evolution equation for the dislocation density (Equation 3) represents a competition between storage of mobile dislocations and dynamic annihilation of dislocations. The recovery is simply modelled by a dynamic recovery term, i.e. increasing with the viscoplastic strain rate and equal to zero without viscoplastic strain rate. This point is further discussed in Reference [7].

While the parameter k_1 depends on temperature only, k_2 depends on both temperature and strain rate. To simplify the procedure of determining these parameters by comparing modelled curves to tensile test data, the well-known Zener-Hollomon [14] or Garofalo's equation [15] is assumed to apply in the so-called saturation stress limit, $\frac{\partial \rho}{\partial t} = 0$, in which the dislocations are recovered at the same rate as they are created. Moreover, the Zener-Hollomon relation provides useful empirical information on the saturation stress at high temperatures, which complete the experimental data given by the somewhat limited Gleeble test data to which the model has been fitted (see below).

The precipitation model (Equations 4 and 5) describes the nucleation, growth and coarsening of the precipitates, and the evolution process is controlled by the solute concentration and consequently by the temperature. The particles are assumed spherical and with uniform thermodynamic properties. From the evolution of the distribution of particles size and density, the strengthening effects are deduced. The precipitate stress is calculated from the particle density, the mean radius, and the mean strength of the particles (Equation 8).

In the present model, the primary variable is the total strain rate which includes the elastic strain rate (Hookes law) and the thermal expansion (Equations 9 to 11). This general formulation allows the integration of the constitutive equations for any evolutions of temperature and total strain rate.

The parameters of Equations 1 to 3 have been determined by fitting calculated tensile stress-strain curves at constant temperatures and strain rates to similar experimental curves for a 6060 aluminium alloy established by means of a Gleeble 3500 R thermo-mechanical simulator. These mechanical tests were carried out for temperatures and strain rates in the range 20 to 340 °C and 10^{-4} to 10^{-2} s^{-1}, respectively. The test specimens were solution heat treated, water quenched, and artificially aged for 9 hours at 170 °C to reach the peak-aged (T6) temper condition. The samples were first strained at 6 % at a constant temperature. A rapid heating cycle was then imposed with a peak temperature of 540 °C, followed by an additional 8 to 10 % straining at the same temperature. In both stages, after a first 2 to 4 % straining, the strain rate was increased by a factor 10 from a value of about 10^{-3} s^{-1}. Again, the reader should consult Reference 7 for details. The parameters of Equations 4 to 8 were taken from Reference 11.

The hardening curve approach

Hardening curves have been established by simulating a series of tensile tests using the new constitutive model. These simulations were carried out for seven constant temperatures (20°C; 100°C; 200°C; 300°C; 400°C; 450°C; 500°C) and three constant values of the total strain rate ($10^{-3}s^{-1}$; $10^{-2}s^{-1}$; $10^{-1}s^{-1}$). The corresponding 21 stress-strain curves are then assumed to characterise the material behaviour for the hardening curve approach. During metal forming simulation using this approach, the current value of the stress is obtained by direct interpolation between these 21 curves or by an extrapolation from the closest points if the temperature or strain rate turns out to be outside the mapped interval. We emphasise that the same model is used for generating the stress-strain curves for the hardening curve

Table 1: Main equations of the constitutive model

$$\sigma = \chi\left(T\right)\hat{\sigma}\left(\frac{\dot{\epsilon}^p}{\dot{\epsilon}_0}\right)^{1/m} \tag{1}$$

$$\hat{\sigma} = \sigma_\perp + \sigma_{ppt} = M\alpha Gb\sqrt{\rho} + \sigma_{ppt} \tag{2}$$

$$\frac{\partial \rho}{\partial t} = \left[k_1\sqrt{\rho} - k_2\,\rho\right]\dot{\epsilon}^p \tag{3}$$

$$j = j_0\,exp\left[-\left(\frac{A_0}{RT}\right)^3\left(\frac{1}{ln\left(\bar{C}/C_e\right)}\right)^2\right]\,exp\left(-\frac{Q_d}{RT}\right) \tag{4}$$

$$\frac{dr}{dt} = \frac{\bar{C} - C_i}{C_p - C_i}\frac{D}{r} \tag{5}$$

$$\frac{\partial N_v\left(r,t\right)}{\partial t} = -\frac{\partial}{\partial r}\left[N_v\left(r,t\right)\frac{dr}{dt}\right] + j \tag{6}$$

$$\bar{C} = C_0 - \left(C_p - \bar{C}\right)\int_0^\infty \frac{4}{3}\pi r^3\varphi dr \tag{7}$$

$$\sigma_{ppt} = \frac{M}{b^2\sqrt{G}}\sqrt{\frac{N_v\bar{r}}{\beta}}\,\bar{F}^{3/2} \tag{8}$$

$$\dot{\epsilon} = \dot{\epsilon}^e + \dot{\epsilon}^p + \dot{\epsilon}^T \tag{9}$$

$$\dot{\epsilon}^T = \alpha_T\frac{\partial T}{\partial t} \tag{10}$$

$$\sigma = E\,\epsilon^e \tag{11}$$

approach as that being used in the simulation based on the new internal variable model. This fact rules out any possible differences between the hardening curves and the experimental curves to which the new internal variable model has been tuned.

The hardening curves for 20, 300, and 450°C are displayed in Figure 2. While the strain rate sensitivity is nearly negligible, and the hardening is nearly linear at 20°C, the strain rate effect sensitivity is strong, and the saturation stress is quickly reached at 450°C. For both these "extreme" temperatures, the precipitate state remains constant during the tensile test. At the "intermediate" temperature of 300°C however, the stress evolution at the lowest strain rate is more complex because of the combined effects of a significant work hardening and a non-negligible evolution of the precipitates size and density. More specifically; the particles have time to nucleate and grow in this case.

Case studies

Process description

The forming operation of imprints in plates or profiles is traditionally performed by the means of a tool and a die or backing tool. However, when long extruded sections are deformed, it can be quite complicated to insert the backing tool inside the work piece. This limitation can be dealt with by the use of the local hot forming process [2] in which the selected region is rapidly heated locally. The softening of the material at the elevated temperatures then allows the forming operation to be performed by a simple tool as the surrounding stronger material acts as a die.

Table 2: Nomenclature

Variables		Unit
σ	Stress	MPa
$\sigma_\perp$	Dislocation stress	MPa
σ_{ppt}	Precipitate stress	MPa
$\hat{\sigma}$	Hardening stress	MPa
$\dot{\epsilon}$	Total strain rate	s^{-1}
$\dot{\epsilon}^p$	Viscoplastic strain rate	s^{-1}
$\dot{\epsilon}^e$	Elastic strain rate	s^{-1}
$\dot{\epsilon}^T$	Thermal strain rate	s^{-1}
ρ	Dislocation density	m^{-2}
N_v	Precipitate density	$\#/m^3$
$\bar{F}$	Mean obstacle strength	MPa/m
T	Temperature	K
$\bar{C}$	Mean solute concentration	$wt\%$
C_e	Equilibrium solute concentration	$wt\%$
C_p	Particle solute concentration	$wt\%$
C_i	Interface solute concentration	$wt\%$
α_T	Thermal dilatation coefficient	K^{-1}
r	Particle radius	m
r^*	Nucleation radius	m
φ	Size distribution function	m^{-4}
Parameters		Value
M	Average taylor factor	3.06
E	Young's modulus	Temperature dependant (see Ref. 7)
G	Shear modulus	Temperature dependant (see Ref. 7)
b	Burgers vector	$2.86 \cdot 10^{-10}\ m$
α	Numerical constant	0.3
$\dot{\epsilon}_0$	Reference strain rate	$60000\ s^{-1}$
m	Strain rate sensitivity	Temperature dependant (see Ref. 7)
k_1	Dislocation storage coefficient	$3.5 \cdot 10^8\ m^{-1}$
k_2	Dynamic recovery coefficient	From Zener-Hollomon relation (see Ref. 7)
χ	Temperature sensitivity	Temperature dependant (see Ref. 7)
D	Solute diffusivity	$2.2 \cdot 10^{-4}\ m^2/s$
j_0	Reference nucleation rate	$3.07 \cdot 10^{36}\ \#/m^3/s$
A_0	Energy barrier for nucleation	$18\ kJ/mol$
Q_d	Diffusion activation energy	$130\ kJ/mol$
β	Numerical constant	0.53
r_c	Critical radius	$5.7 \cdot 10^{-9}\ m$

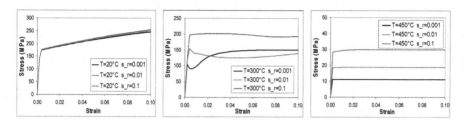

Figure 2: Stress-strain curves at 20°C, 300°C and 450°C, and different strain rates (s_r)

Figure 3 indicates the extruded section and the tool along with the temperature field at a given instant during cooling. The induction coil has been omitted from the figure. The highest temperatures are just below the tool where the deformation is to take place.

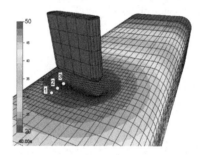

Figure 3: Sketch of the tool and local heating of the extruded part. The three locations of the typical temperature and strain rate evolutions shown in Figure 4 are also indicated.

Basis for the comparison: representative temperature and strain rate evolutions

In order to specifically focus on comparing the constitutive models, we have established three sets of representative, and typical, evolutions of temperature and strain rate development at different locations of the work piece. These evolutions are then used as input to calculations of the flow stress evolution by means of the two constitutive approaches; i.e., the hardening curve approach and the new constitutive model integrating the equations in Table 1. Please note that in the latter approach, the equations are integrated implicitly at each time step, and since the precipitate evolution depends on temperature only, its governing equations can be decoupled and solved separately.

To establish the three representative temperature and strain rate evolutions, the FE modelling software WELDSIM described elsewhere [17] was first applied. Smoothed and simplified versions of the evolutions in Locations 1, 2, and 3 indicated in Figure 3 were then made (see Figure 4). These are typical for the low, intermediate and high temperature regimes of the work piece.

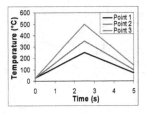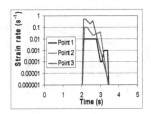

Figure 4: Representative histories of temperature and strain rate at Location 1, 2, and 3 indicated in Figure 3.

Results and discussion

The left graph in Figures 5, 6, and 7 shows the evolution of the flow stress for the three different evolutions of temperature and strain rate shown in Figure 4, respectively, calculated by the two approaches. While the heating stage before the deformation starts is not displayed, the cooling stage is included because the internal variables can continue to evolve during this stage. Furthermore, the middle and right graphs of these three figures display the precipitate stress (σ_{ppt}) and dislocation stress ($\sigma_\perp$) (see Equation 2). These two quantities are well defined in the internal variable model; however, their introduction in the hardening curve approach requires a more thorough explanation. For the calculation of the hardening curves using the constitutive model, the quantities $\sigma_\perp$ and σ_{ppt} were calculated. These quantities thus represent the microstructure at these instants, and indicate the different contributions to the flow stress. Please note however, that this information is not available in a "real" situation of using empirical hardening curves as the basis for the constitutive in the FE simulation. The evolutions of $\sigma_\perp$ and σ_{ppt} for the hardening curve approach have been displayed in Figures 5 to 7 only to highlight and explain the differences between the two approaches.

In the first case (Figures 5), the material experiences relatively low temperatures and strain rates. The work hardening is then low and the precipitate stress remains constant. This leads to relatively similar evolutions of the flow stress in the two approaches. The small discrepancies between the curves stems mainly from the fact that the dislocation density evolution is not taken into account in the hardening curve approach. At a given strain, the flow stress predicted by the hardening curve approach corresponds to a tensile test performed at the instantaneous strain rate. But the material might have experienced other strain rates, and then the dislocation density can be quite different, as between 2.5 and 3 s. In addition, there is inaccuracy associated with the applied interpolation (e.g. overestimated stress at the end).

The second case (Figures 6) shows another aspect related to the evolution of the precipitate stress during the process. The flow stresses calculated by the two approaches have a similar evolution during the first part of the forming operation, even though the faster precipitate dissolution predicted by the hardening curves approach leads to lower stress levels. However, after 2.6 s, the precipitates have been partially dissolved and there is not enough time to nucleate/growth new. This means that the precipitate stress remains low and constant after 2.6 s. The hardening curves, on the other hand, reflect the evolution of the initial microstructure at the current temperature, and consequently, the partial dissolution of the precipitates is thus ignored when low temperatures are reached. Because if the deformation

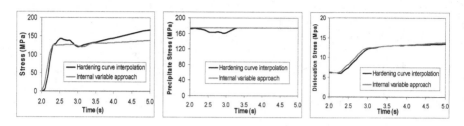

Figure 5: Flow stress, precipitate stress, and dislocation stress evolutions calculated by the two constitutive approaches using the temperature and strain rate at Location 1 as input.

was made at these low temperatures, the precipitates would not have been dissolved. The material keeps its original strength in this model approach. Furthermore, the evolution of the dislocation density (and thus the dislocation stress) is quite irregular and does never reach the saturation plateau. In summary, these two effects results in a far too high flow stress (204 MPa instead of 58 MPa) by the end of the process.

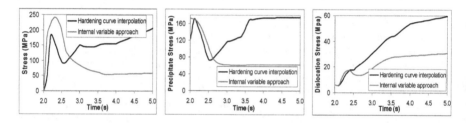

Figure 6: Flow stress, precipitate stress, and dislocation stress evolutions calculated by the two constitutive approaches using the temperature and strain rate at Location 2 as input.

In the third case (Figure 7) the high temperature regime quickly dissolves all the precipitates, and these are only to a small extend able to nucleate and grow again during the cooling stage. The precipitate stress associated with the hardening curve approach reaches the complete dissolution state, but after 3.4 s it goes back to the initial state due to the decrease in temperature. The precipitate history is thus not taken into account: the microstructure is represented by a precipitate stress of 170 MPa instead of 27 MPa obtained by the internal variable approach. Moreover, the precipitate stress curve has a non-physical discontinuity due to the "jump" between the hardening curves. The evolution of the dislocation density is also critical. Even though the dislocation stress curves look quite similar before 2.5 s, their differences have a pronounced effect upon the final stress curve. For example, the overestimation of the dislocation density at the beginning of the deformation leads to a peak flow stress of 83 MPa in the hardening curve approach, while the internal variable approach predicts 41 MPa. Furthermore, during the cooling stage, the saturation plateau is not obtained for the hardening curve approach. This leads to a very strong hardening. In conclusion, the combined effects of the precipitate and dislocation evolution leads to a significant gap in the predicted stress at the location experiencing the larger deformation, and thus defines the final shape of the imprint made in the hot forming process.

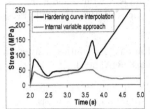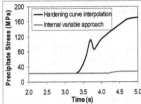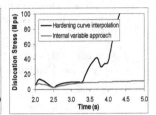

Figure 7: Flow stress, precipitate stress, and dislocation stress evolutions calculated by the two constitutive approaches using the temperature and strain rate at Location 3 as input.

Conclusion and further research

An internal variable constitutive model for the use in FE simulations of local hot forming of 6xxx alloys has been briefly presented. The model relates the flow stress during viscoplastic deformation to the temperature and strain rate, and accounts for the accumulation and annihilation of dislocations, the changes in the volume fraction and size distribution of the hardening precipitates, and how these phenomena influence the flow stress.

The present study has shown that the evolution of precipitates and dislocation density can have a significant effect upon the flow stress during local hot forming operations and that these effects should be accounted for in FE simulations of the process. If instead a more traditional approach based on interpolation between a set of hardening curves is applied, the predicted flow stress might be wrong and thus leading to a failure in an optimisation procedure. This lack of reliability applies even when the stress-strain curves quite accurately quantify the flow stress for the conditions applying in the mechanical test carried out to establish the hardening curves.

Acknowledgement

The authors would like to thank Professor Ø. Grong (NTNU, Norway) and Dr. K.O. Pedersen (SINTEF, Norway) for fruitful discussions and for their contribution in the development of the constitutive model summarized in Reference 7. Financial support from the Research Council of Norway through Project No. 158918/130 *Computer Aided Local Thermal Manipulation of Aluminium Alloys* managed by the Institute for Energy Technology, Norway, is greatly acknowledged.

References

[1] B.W. Tveiten, A. Fjeldstad, G. Härkegård, O.R. Myhr, and B. Bjørneklett. Fatigue life enhancement of aluminium joints through mechanical and thermal prestressing. *Int. J. of Fatigue*, In press, 2006.

[2] B. Bjørneklett, O.E. Myhr, and P. Vist. Local forming, locally formed work piece and tool for such forming. *Patent application WO 2005/077560 A1*, August 2005.

[3] U.F. Kocks. Laws for work-hardening and low-temperature creep. *J. Eng. Mat. Tech.*, pages 76–85, 1976.

[4] Y. Estrin. Dislocation-density-realted constitutive model. In A.S. Krausz and K. Krausz, editors, *Unified constitutive laws of plastic deformation*, pages 69–106. Academic Press, 1996.

[5] X. Wang, W.J. Pool, S. Esmaeili, D.J. Lloyd, and J.D. Embury. Precipitation strengthening of the Aluminum Alloy AA6111. *Metall. Mat. Trans.*, 34A, 2003.

[6] LM. Cheng, W.J. Poole, J.D. Embury, and D.J. Lloyd. The influence of precipitation on the work-hardening behavior of the aluminum alloys AA6111 and AA7030. *Met. Mat. Trans.*, 34A, 2003.

[7] S. Gouttebroze, A. Mo, Ø. Grong, K.O. Pedersen, and H.G. Fjær. A new constitutive model for the FE simulation of local thermal forming of aluminium 6xxx alloys. *(to be submitted)*, 2006.

[8] *ABAQUS v6.5, Analysis User's Manual, Section 11.2.3.*

[9] A. Deschamps, D. Solas, and Y. Bréchet. Modeling of microstructure evolution and mechanical properties in age-hardening aluminum alloys. In Y. Bréchet, editor, *Microstructure Properties and Processes, Proceedings of Euromat 99, München, Germany*, volume 3, pages 121–132. Wiley-VCH, 1999.

[10] O.R. Myhr, Ø. Grong, and S.J. Andersen. Modelling of the age hardening behavior of Al-Mg-Si alloys. *Acta Mater.*, 49, 2001.

[11] O.R. Myhr, Ø. Grong, H.G. Fjær, and C.D. Marioara. Modelling of the microstructure and strenght evolution in Al-Mg-Si alloys during multistage thermal processing. *Acta Mater.*, 52, 2004.

[12] A. Deschamps and Y. Bréchet. Influence of predeformation and ageing of an Al-Zn-Mg alloy - II. modeling of precipitation kinetics and yield stress. *Acta Mater.*, 47:293–305, 1999.

[13] S. Esmaeili, D.J. Lloyd, and W.J. Poole. A yield strength model for the Al-Mg-Si-Cu alloy AA6111. *Acta Mater.*, 51, 2003.

[14] C. Zener and J.H. Hollomon. Effect of strain rate upon plastic flow of steel. *J. Appl. Phys.*, 15(1), 1944.

[15] F. Garofalo. *Fundamentals of creep and creep-rupture in metals.* The MacMillan Company, New-York, 1965.

[16] A.H. Van den Boogaard and J. Huétink. Simulation of aluminium sheet forming at elevated temperatures. *Comput. Methods Appl. Mech. Engrg.*, 2006.

[17] H.G. Fjær, B.I. Bjørneklett, and O.R. Myhr. Microstructure based modelling of Al-Mg-Si alloys in development of a local heating processes for automotive structures. In *Proceedings of the TMS Annual Meeting, San Fransico, USA*, pages 95–100, February 2005.

THE INFLUENCE OF PARTICLE SIZE TO THE PREPARATION OF FOAM ALUMINUM BY POWDER METALLURGY METHOD

Zhiqiang Guo[1], Guangchun Yao[2], Yihan Liu[3]

[1]Zhiqiang Guo (School of Materials&Metallurgy, Northeastern University);
Shenyang, Liaoning 110004, China
[2]Guangchun Yao Professor, E-mail: G.C.Yao@mail.neu.edu.cn
Shenyang, Liaoning 110004, China
[3]Yihan Liu Associate Professor of Northeastern University

Keywords: powder metallurgy, foam aluminum, particle size, TiH_2.

Abstract

There are many factors influence th e preparation of foam aluminum. In this paper influences of particle sizes to powder m etallurgy method were studied. Three different sizes of powders were applied: d<0.074mm ; 0.074mm <d<0.0.147mm; 0.147 mm<d<0.295mm. Mixed powders with foaming agent (TiH_2) respec tively, pressed th em into precursors. And then did foa ming experiments. It can be conclude d with the increasing of powder diameter, reunion phenom ena of TiH_2 became serious. TiH_2 particles reunited together. In foaming process, some places that TiH_2 reunited expanded violently. A po rtion of pore walls becam e th inner fleetly because of the shortage of resilience, pore walls began to rupt ure, collapse occurred. The results s howed that high porosity and stability foam alum inum m aterials can be obtained by   using particles wit h small diameter (d<0.074mm).

Introduction

Foam aluminum is a new functional m aterial that has been developed in recent years. It attaches more attentions, [1,2] because of its super physical propertie s and special m echanical properties. There are m any ways to produce foa   m aluminum. Powder m etallurgy m ethod that can manufacture fairly inex pensive clo sed cell ma terials with attractive m echanical prope rties became one of the popular ways.[3-6] Fraunhofer-institute for advanced materials of Germany had a deep research on that m ethod,[7,8] alum inum foam sandwi ch pane ls are produced and used to car coping s, som e se mi-manufactured accesso ries were also produced. In china research es on powder m etallurgy m ethod were just at its starting period. [9] All rese arches were c arried out in laboratories. Further more, major works were concentrated on capabilities of foam aluminum.

In this p aper, particle sizes tha t influence the preparation of foa m aluminums were studied. The distributing trends of TiH 2 particles in precursors and influe nces of hot-pressing to precursors were also obtained. Though com paring to foam ing effect of precursors which were m ade by alloy powders of different particle sizes, optimal particle size of alloy powder was confirmed.

Process of the experiment

In this paper Al-Si alloy powders are used as the raw material, alloy powders with different particle size are applied: d<0.074mm; 0.074mm<d<0.0.147mm; 0.147mm<d<0.295mm. Mix the alloy powder with foaming agent (TiH_2) that the amount is 0.6%, about 10 hours on electric mixing machine. 100g alloy powders are pressed into a precursor with a diameter of 50mm under the pressure of 300MPa by cold pressing + hot pressing way. Put the precursor into the experimental set-up for 1 hour at 400°C, and then do the foaming experiment, foam aluminum materials are obtained. The precursors are analyzed by SEM. Photographs of foam aluminums are scanned by a scanner. Though analyzing to SEM photos of precursors and photos of foam aluminum, influences of particle sizes to the preparation of foam aluminum are obtained. Fig.1 shows the major experimental set-up for foaming tests.

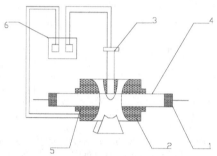

Fig. 1 Schematic of experimental set-up for foaming tests
1, rubber plug 2, heat preservation layer
3, thermocouple 4, hearth 5, globar resistance heater 6, temperature control equip

The results and the analyses

<u>Influences of particle sizes to the distributing of TiH_2</u>

In this paper, three kinds of alloy powders with different particle sizes are applied: d<0.074mm; 0.074mm<d<0.0147mm; 0.147 mm<d<0.295mm. Fig 2 shows the distributing of foaming agent (TiH_2) in precursors. It can be shown with the decreasing of particle sizes of alloy powders,

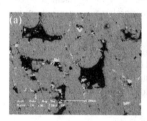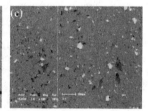

Fig 2 Distributing of TiH2 particles in precursors of different particle sizes before hot-pressing
(a) 0.147 mm<d<0.295mm; (b) 0.074mm<d<0.0.147mm; (c) d<0.074mm

the distributing of foaming agent (TiH₂) changed gradually from "line-conglomeration" forms to "spot" forms. The foaming agent (TiH₂) used in this experiment are powders of d<0.074mm. When the granularities of alloy powders were 0.147 mm<d<0.295mm, alloy powders were much bigger than TiH₂ particles, so there were much space among alloy powders, TiH₂ particles reunited together in the space of several alloy particles, just like a "conglomeration", and it showed a "line" form in the border of two alloy powders. So it showed a distributing of "line-conglomeration" forms. With the decreasing of alloy powders granularities, "line-conglomeration" phenomena also decreased. But when the particles sizes of alloy powders decreased to 0.074mm<d<0.0.147mm, alloy powders were still too big to avoid the appearance of "line-conglomeration" phenomena, they were just a little weaker than 0.147 mm<d<0.295mm. When the particles sizes of alloy powders decreased to d<0.074mm, alloy powders and TiH₂ powders reached the same granularities level, "line-conglomeration" phenomena disappeared, TiH₂ powders showed "spot" forms equally distribute in precursors.

<u>Influences of hot pressing</u>

The densification of the precursor that can foam must exceed 96%,[10] however, because of the existence of "arch bridge" phenomenon, there were much space existed in alloy powders, so after cold pressing only 80% densification can be achieved. Therefore cold pressing + hot pressing way was used in this work.

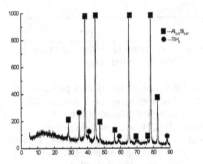

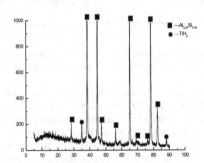

Fig 3 XRD pattern for metal
matrix before hot-pressing

Fig 4 XRD pattern for metal
matrix after hot-pressing

Fig 3 and fig 4 showed XRD pattern for metal precursors before and after hot pressing. It can be shown that the composition of precursors had no changes before and after hot pressing. They were still $Al_{3.21}Si_{0.47}$ alloy and foaming agent TiH₂. Because an oxide layer formed at surface in heat treatment process, it stopped the interior part of precursor from being oxidized, so hot pressing had no effect to the composition of precursor.

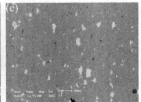

Fig 5 Distributing of TiH$_2$ particles in precursors of different particle sizes after hot-pressing
(a) 0.147 mm<d<0.295mm; (b) 0.074mm<d<0.0.147mm; (c) d<0.074mm

Fig5 showed the superficial appearance of precursor and the distributing of TiH$_2$ particles after hot pressing. It can be shown after hot pressing "arch bridge" phenomenon disappeared, so the space in alloy powders also disappeared. The theory foaming densification (overrun 96%) was obtained. The distributing of TiH$_2$ particles after hot pressing has no changing: with the decreasing of particle sizes of alloy powders, the distributing of foaming agent (TiH$_2$) changed gradually from "line-conglomeration" forms to "spot" forms. It just with the increasing of densification, the distributing of TiH$_2$ particles became more tightness. So hot pressing had no influences to composition of precursors and distributing of TiH$_2$ particles. It just improved the densification to reach the intention of foaming.

Foaming effect of precursor

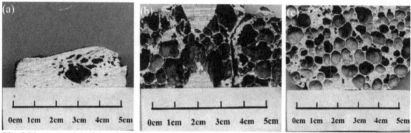

Fig 6 Photographs of foam aluminum that made by alloy powders with different particle sizes
(a) 0.147 mm<d<0.295mm; (b) 0.074mm<d<0.0.147mm; (c) d<0.074mm

Fig 6 showed the foaming effects of precursors. When the granularities of alloy powders were 0.147 mm<d<0.295mm, because of the distributing of "line-conglomeration" forms, TiH$_2$ particles reunited too serious, in foaming process, H$_2$ was largely emitted in "line-conglomeration" areas of TiH$_2$, the oxide layer was smashed, H$_2$ was lost from surface, and in large area, the contents of TiH$_2$ were too low to foam, so the precursor just had a little expansion. With the decreasing of particle sizes (0.074mm<d<0.0.147mm) the foaming effect improved markedly, however, particle sizes were still too big, so "line-conglomeration" phenomenon still existed in precursor. Therefore the foam still had no regular shape, and distributing un-uniformity, some area foams un-markedly. When particle sizes decreased to d<0.074mm, foaming agents (TiH$_2$) equably distribute in precursor. H$_2$ was emitted placidly. So the foam aluminum prepared in this condition had thin cell walls, equably distributing, prefect growth.

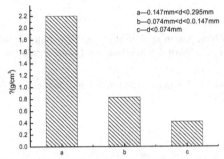

Fig 7 Densities of foam aluminum which were prepared
by alloy powders with different particle sizes

Fig 7 shows the densities of foam aluminum. With the decreasing of particle sizes of alloy powders, "line-conglomeration" phenomenon gradually decreases, the densities of foam aluminum also decrease. When particle sizes decrease to d<0.074mm, the density of foam aluminum reaches the mini-value $0.42g/cm^3$. High quality foam aluminum with low density, high porosity, thin cell walls and perfect growth are prepared in this condition.

Conclusion

(1) When the particle sizes is much bigger than TiH_2 particles, TiH_2 particles presenting "line-conglomeration" forms distribute in the space among alloy powders. With the decreasing of particle sizes of alloy powders, the distributing of foaming agent (TiH_2) change gradually from "line-conglomeration" forms to "spot" forms. When the particles sizes of alloy powders decrease to d<0.074mm, alloy powders and TiH_2 powders reached the same granularities level, "line-conglomeration" phenomena disappeared, TiH_2 powders show "spot" forms equally distribute in precursors.

(2) Hot pre ssing had n o inf luences to com position of pr ecursors an d distribu ting of T iH$_2$ particles. It just im proved the densification, to reach the in tention of foam ing. W hen particle sizes decrease to d<0.07 4mm, the density of foam aluminum reaches the m ini-value 0.42g/cm3. High quality foam aluminum with low density, high porosity, thin cell walls and perfect growth are prepared.

Reference

[1] Davies G J, Zhen S. "Metallic foams: their production, Properties and application". *Material Science*, 1983, 18(4), 1899-1903.
[2] Wang Zhutang. "Aluminium foams: production processes, structure and properties, applications and market(3)". *Light Alloy Fabrication Technology*, 1999, 27(12):1-5.
[3] Stanzick H, Wichmann M, Banhart J. "Process control in aluminum foam production use real-time X-ray radioscopy". *Advanced Engineering Materials*, 2002, 4(10): 814-823.
[4] Banhart J, Stanzick H. "Metal foam evolution studied by synchrotron radioscopy". *Applied hysics Letters*, 2001, 78(8):1152-1154.
[5] Baumeister J, Babhart J, "Weber M. Aluminum foams for transport industry". *Materials&Desgn*, 1997, 18(4):217-220.
[6] Duarte I, Weigand P, Banhart J. "Foaming kinetics of aluminum alloys". Banhart J, Ashby M F, leck N A. "Metal Foam sand Porous Metal Structure". Bremen: MIT Verlag, 1999. 97-104.
[7] Fuganti A, Lorenzi L, Hanssen A, et al. "Aluminum foam for automotive applications". *Advanced Engineering Materials*, 2000, 2(4): 200-204.
[8] John Banhart, "Metal foams near commercialization ". *PM Special Feature*. 1997, 41(4): 38-41.

[9] Wei Li, YAO Guangchun, Zhang Xiaoming, Luo Hongjie. "Preparation of foam aluminum by powder metallurgy process". *Journal of Northeastern University (Natural Science)*, 2003, 24(11):1071-1074。

[10] Duarte I, Banhart J. "A study of aluminum foam formation—kinetics and microstructure". *Acta Mater*, 2000, 48(2): 2349-2362.

Aluminum Alloys for Transportation, Packaging, Aeropsace, and Other Applications
Edited by Subodh K. Das, Weimin Yin
TMS (The Minerals, Metals & Materials Society), 2007

FRICTION STIR WELDING OF ALUMINUM ALLOYS 6061-T6 AND 6101-T6

Dr. Carter Hamilton[1], Dr. Stanisław Dymek[2]

[1]Department of Mechanical and Manufacturing Engineering
Miami University, Oxford, OH

[2]Faculty of Metals Engineering and Industrial Computer Science
AGH University of Science and Technology, Kraków, Poland

Keywords: friction stir welding, aluminum, microstructure, mechanical properties, material flow

Abstract

Tin plated 6061-T6 and 6101-T6 aluminum extrusions were friction stir welded in a corner configuration. A banded microstructure of interleaved layers of particle-rich and particle-poor material comprised the weld nugget. Transmission electron microscopy revealed the strong presence of tin within the particle-rich bands, but TEM foils taken from the TMAZ, HAZ and base material showed no indication of Sn-containing phases. Since tin is limited to the surface of the pre-weld extrusions, surface material flowed into the nugget region, forming the particle-rich bands. Similarly, the particle-poor bands with no tin originated from within the thickness of the extrusions. Mechanical testing of specimens excised from the as-welded panels resulted in consistent failure on the retreating side of the weld, approximately 10 mm from the center of the weld nugget. SEM analysis of the fracture surfaces revealed typical ductile rupture; however, when the excised specimens were solution heat treated and aged, the specimens consistently failed within the weld nugget and revealed two distinct modes of fracture characteristic to the particle-rich and particle-poor layers of the nugget. This investigation correlates the mechanical performance of friction stir welded 6061-T6 and 6101-T6 panels with the fracture behavior and the microstructural characteristics of the weld nugget. Factors leading to the development of the banded microstructure are identified.

Introduction

Invented in 1991 by The Welding Institute, Friction Stir Welding (FSW) is a novel solid-state joining process that is gaining popularity in the manufacturing sector [1, 2]. FSW utilizes a rotating tool design to induce plastic flow in the base metals that essentially "stirs" the pieces together. During the welding process, a pin, attached to the primary tool, is inserted into the joint with the shoulder of the rotating tool abutting the base metals. As the tool traverses the joint, the rotation of the shoulder under the influence of an applied load heats the metal surrounding the joint and with the rotating action of the pin induces metal from each workpiece to flow and form the weld. The microstructure resulting from the influence of plastic deformation and elevated temperature is characterized by a central weld nugget surrounded by a thermo-mechanically affected zone (TMAZ) and heat affected zone (HAZ). The welded joint is fundamentally defect-free and displays excellent mechanical properties when compared to liquid-state welds [3, 4, 5]. Since no melting occurs during FSW, the process is performed at much lower temperatures than conventional welding techniques and circumvents many of the environmental and safety issues associated with these welding methods. Due to the associated benefits, the aerospace industry is embracing FSW technology and implementing new FSW capabilities.

Over the last fifteen years, numerous investigations have sought to characterize the principles of FSW and to model the microstructural evolution. The current status of FSW research has been well summarized by Mishra and Ma [6]. The flow of material during FSW is a complex process that is not fully understood despite these investigations and models. Several studies have compared material flow during FSW with wrought metal processes and have modeled weld nugget development as an extrusion process [6, 7]. In particular, Krishnan and Sutton et al. hypothesized that the nugget forms as a volume of material from the weld surface extrudes into the joint during each revolution of the tool [8, 9]. Seidel and Reynolds utilized fluid mechanics to create a two-dimensional model of FSW as a non-Newtonian fluid flowing around a rotating cylinder [10]. Though each of these models predict some characteristics of the weld nugget, there are limitations. For example, the models do not adequately describe the formation of the banded microstructure of particle-rich and particle-poor regions commonly observed during FSW [8, 11]. The following study investigates friction stir weld nugget development during the welding of 6101-T6 and 6061-T6 extrusions that were plated with tin to facilitate the tracking of material flow.

Experimental Procedure

<u>Friction Stir Welding</u>
Aluminum 6061-T6 and 6101-T6 extrusions produced in accordance with ASTM B 317 with a thickness of 6.35 mm and a width of 195.0 mm were obtained and welded in the corner configuration represented in Figure 1 [12]. As shown in the diagram, with the clockwise tool rotation, FSW occurs along the L-direction of the advancing side (rotation of the tool is in the same direction as the weld direction) and along the LT-direction of the retreating side (rotation of the tool is in the opposite direction of the weld direction). Prior to welding, the extrusions were plated with tin, 0.05 mm thick, in order to trace the flow of surface material during the process. The diameter of the FSW tool shoulder was 21.0 mm, the pin diameter was 6.5 mm and the pin depth was 5.5 mm. More specific details of the tool design are proprietary to the welding company, but Mishra and Ma have reviewed many of the common FSW tool designs, which are also representative of those utilized in this investigation [6, 13, 14]. The extrusions were friction stir welded with a tool rotation speed of 900 rpm, a traverse speed in the weld direction of 5.0 mm/s and an applied force of 22.5 kN.

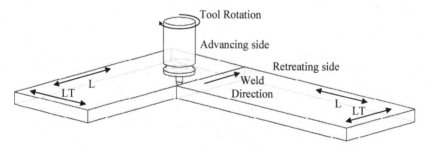

Figure 1 - Schematic of corner weld geometry and FSW orientation

Tensile Specimens and Mechanical Testing

For mechanical testing, full thickness (6.35 mm) tensile samples were excised from the corner welds as shown in the plan view of Figure 2. With this specimen orientation, the weld is centered along the tensile specimen, and as such, the load is applied transverse to the weld direction and across all microstructural regions associated with the welding process, i.e. weld nugget, HAZ and TMAZ. The geometry and dimensions of the welded tensile specimens are also shown in Figure 2 with the longitudinal direction indicated. During tensile testing, therefore, the tensile load is applied in the LT-direction along one side of the specimen (the advancing side during FSW) and in the L-direction along the other side (the retreating side during FSW). In addition to the welded tensile specimens, tensile bars of the same geometry and dimensions were also excised from an area well away from the weld region for baseline property comparison.

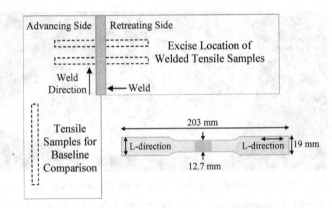

Figure 2 - Excise location of tensile specimens from the corner weld configuration

All tensile tests were performed in accordance with ASTM E 8 utilizing an Instron 5867 screw driven test frame with a 30 kN load cell and a 0.001 – 500 mm/min speed range [15]. Specimen extension, crosshead deflection and load were recorded throughout the test duration. Specimen extension was measured by means of a 25.4 mm extensometer attached to the reduced section that spanned the width of the weld. The extensometer remained attached to the specimen through yielding, but was removed prior to specimen failure to prevent damage to the equipment. The yield stress, σ_y, was obtained by the 0.2% offset method, and the elastic modulus, E, was determined by fitting a linear regression to the elastic region of the stress-strain curve. Elongation was determined by the extent of crosshead deflection relative to the initial length of the specimen between the machine grips and by scribing marks 50.8 mm apart within the reduced section prior to testing and measuring their separation after testing.

Post-Weld Heat Treatments

To assess the impact of post-weld heat treatments on the mechanical behavior of friction stir welded sections, select tensile specimens excised from the welded 6061-T6 and 6101-T6 extrusions received additional heat treatment. The soak temperatures and times were chosen in accordance with the Aluminum Standards and Data for solution heat treatment and aging of 6XXX series alloys [16]. Welded tensile specimens of 6061-T6 received post-weld aging at 200°C for 10 hours, and welded tensile specimens of 6101-T6 received post-weld solution heat treatment at 510°C, followed by a water-quench and subsequent aging at 200°C for 10 hours. After heat treatment, all tensile specimens were tested in accordance with ASTM E 8 utilizing the same instrumentation detailed above.

Results and Discussion

<u>Evolution of the Weld Nugget</u>
Figure 3 shows the microstructure of a typical FSW nugget produced during this study with the retreating and advancing sides indicated. The complex flow pattern of the FSW process is clearly evident, and the weld nugget is primarily characterized by the banded microstructure. The contrast between the two bands results from an uneven distribution of secondary phase particles. Also evident in the micrograph is the "tail" on the retreating side that extends from the main body of the weld nugget to the weld surface. The morphology of the nugget and tail suggests that surface material is introduced from the retreating side into the plasticized region surrounding the FSW pin. The average distance from the intersection of the tail with the weld surface to the weld centerline is 9.0 mm.

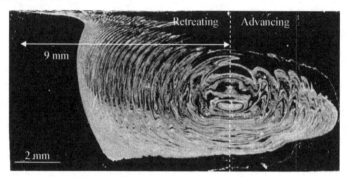

Figure 3 - Representative optical micrograph of friction stir weld nugget

Within the banded microstructure of the nugget, the lighter bands reflect a high density of secondary phase particles, while the darker bands reflect a low one. Since the visual appearance of the bands depends on the instrumentation used, it is more appropriate to differentiate the bands as "particle-rich" and "particle-poor", terminology consistent with that introduced by Sutton et al. who observed the same phenomenon in FSW 2024-T351 rolled sheet [9]. The non-uniform distribution of the secondary phase particles is more evident in the high resolution scanning electron (HRSEM) images presented Figure 4. The image taken from the center of the weld nugget in Figure 4a clearly shows the interleaving of particle-rich and particle-poor layers that develop during FSW. Even more striking, however, is Figure 4b, which shows a very distinct boundary between a particle-rich region of the weld nugget and the particle-deficient TMAZ.

To better analyze the particle distribution and to determine the chemical composition of the particles, themselves, TEM foils were excised from each unique region of the friction stir weld profile, i.e. base material, HAZ, TMAZ and weld nugget. Figure 5a is a TEM micrograph of a foil taken from a region near the center of the weld nugget. The secondary phase particles of the particle-rich bands are clearly evident, and chemical analysis of these phases indicates a strong tin content. Figure 5b is of a foil taken from the TMAZ, the region adjacent to the weld

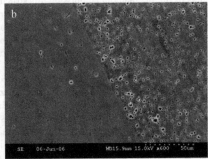

Figure 4 - HRSEM of friction stir weld nugget: a) near nugget center, b) near nugget boundary with TMAZ

nugget. Though the micrograph reveals a high dislocation density in this plastically deformed area, no secondary phase particles are present. In addition, chemical analysis shows no evidence of tin in this microstructural region. Similarly, foils excised from within the HAZ and base material do not indicate the presence of any Sn-containing phases.

Since tin is only present on the surface of the extrusions as plating, the unique occurrence of tin in the weld nugget is extremely significant. Material comprising the weld nugget, therefore, must originate from the extrusion surface since this is the only viable source of tin. Similarly, the material within the TMAZ and HAZ zones must originate from within the extrusion thickness since TEM demonstrated the absence of tin from these regions. The Sn-containing, particle-rich bands necessarily represent surface material that has flowed into the region that ultimately transforms into the weld nugget. Also, since the particle-poor bands do not contain discernable

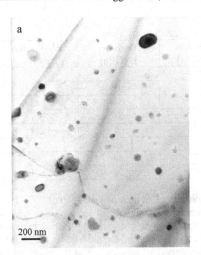

Figure 5 – TEM micrograph of: a) weld nugget center revealing Sn-rich particles, b) TMAZ revealing no Sn

amounts of tin, the material comprising this structure must originate from within the thickness of the extrusions. The banded microstructure, therefore, is composed of interleaved layers of surface material rich in Sn-containing particles and plasticized material already present in the extrusion thickness, i.e. in-situ material. This conclusion supports the hypothesis that the FSW weld nugget forms as surface material extrudes from the retreating side into the plasticized zone.

It should be noted that some researchers have concluded that the weld nugget structure is an interleaving of layers of fine-equiaxed, recrystallized grains with coarse recrystallized grains [17, 18]. Both of these studies, however, examined aluminum alloys other than ones studied in this analysis and utilized different weld configurations and parameters. During this investigation, the grains within the nugget showed little deformation in the flow direction, suggesting that recrystallization did occur; however, the grain size in both the particle-rich and particle-poor bands was equivalent, approximately 10 µm to 20 µm [11].

Mechanical Testing

Table I summarizes the average mechanical properties determined for each material condition of the alloys examined. Typical stress-strain curves are presented in Figure 6. Most notably for the 6101-T6 specimens is the rebound in yield and ultimate tensile strengths from the friction stir weld condition following solution heat treatment and age. Initially, the yield and ultimate tensile strengths decreased by 60% and 35%, respectively, from the baseline condition (in comparison to both the L and LT-directions). Following heat treatment, the yield strength of the 6101-T6 actually rose above the baseline value (16%), and the ultimate strength returned to within statistical equivalence of the baseline value. The elongation of the post-weld heat treatment condition, however, is sharply less than the baseline condition and even less than the elongation realized in the welded condition, reaching only 2.4%. For the 6061-T6 specimens, post-weld aging certainly increased both the yield and ultimate tensile strengths (60% and 12%, respectively) and decreased the elongation. The extent of the changes, however, is not as dramatic for the post-weld aging of 6061-T6, as for the post-weld solution heat treatment and aging of 6101-T6.

Table I - Summary of Mechanical Property Data

Material	Yield Stress σ_y (MPa)	Ten. Stress σ_{TS} (MPa)	Modulus E (GPa)	Elongation e (%)
6101-T6				
Baseline (L)	174	200	65	9.4
Baseline (LT)	172	198	66	14.6
FSW	69	130	64	3.4
FSW + SHT + Age	201	202	64	2.4
6061-T6				
FSW	84	153	60	7.8
FSW + Age	135	172	58	6.2

These results are similar to those observed by Mahoney et al. in their investigation of 7075-T651 plate [6, 17]. During their investigation, the researchers noted a significant decrease in elongation following post-weld heat treatment, as well as an increase in yield strength. In no trial, however, did the yield strength return to or surpass the baseline value, nor did the ultimate tensile strength change from the welded conditions. In addition, the impact of post-weld heat treatment on the mechanical properties was significantly influenced by the grain direction. The authors attributed their observations to an increase in the number of strengthening precipitates and precipitate free zones following post-weld heat treatment.

130

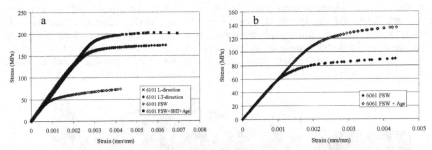

Figure 6- Typical stress-strain curves: a) 6101-T6 and b) 6061-T6

All FSW tensile specimens with no post-weld heat treatments failed on the retreating side of the weld, approximately 10 mm from the center. This distance corresponds approximately to the FSW tool radius of 10.5 mm, thus failure occurred near the transition from the TMAZ to the HAZ on the retreating side. Tensile failure on the retreating side is a common phenomenon, just as other research studies on aluminum alloys 6063-T5 and 7075-T651 have also observed [17, 18]. In these studies, however, the welds were conventional butt welds, and tensile loading was in the LT-direction on both the retreating and advancing sides. In the corner weld configuration of this investigation, the tensile loading on the retreating side was in the L-direction, the strongest grain direction, and on the advancing side, loading was in the LT-direction, the weaker grain direction.

Electron microscopy of the FSW specimens, therefore, examines the fracture surface of the retreating side adjacent to the weld nugget. Figure 7a is an SEM micrograph of a representative FSW fracture surface from the 6101-T6 and 6061-T6 specimens. The fracture surface is primarily characterized by the dimples caused by microvoid nucleation and coalescence typical to ductile rupture and indicative of a high degree of ductility. Very few secondary phase particles are evident on the FSW surface through SEM, though some iron, silicon-containing particles (Fe_xSiAl_y) are seen in the dimples as shown in Figure 7b (approximate particle diameter is 5 μm).

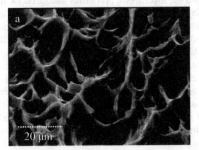 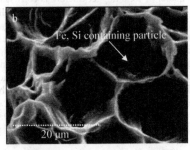

Figure 7 - a) SEM micrograph of representative FSW weld fracture surface showing ductile rupture and b) Higher magnification of same area highlighting iron, silicon containing particle in dimple (Alloy: 6101-T6)

Following post weld heat treatment of the aluminum alloys, the elongation was reduced (most significantly for 6101-T6) and fracture of the tensile specimens occurred within the weld nugget. Examination of these fracture surfaces revealed two distinct fracture morphologies: one characteristic to the particle-rich layer within the weld nugget and the other characteristic to the particle-poor layer. Figure 8 displays a typical fracture surface of the post-weld heat treatment tensile specimens and clearly indicates the two modes of fracture occurring within the weld nugget. On the left of the micrograph, the particle-rich layer reveals microvoid nucleation and coalescence as the primary mode of fracture. EDS analysis of the secondary phase particles within the dimples shows a strong presence of tin, an observation consistent with the TEM results discussed previously. On the right side of the micrograph, the particle-poor layer reveals brittle, cleavage fracture as the primary mode of fracture.

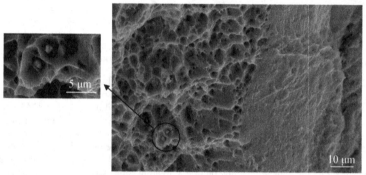

Figure 8 - SEM micrograph showing two modes of fracture within the weld nugget of tensile specimens receiving post-weld heat treatment

Acting independently from each other, these two modes of fracture lead to opposing performances in ductility, microvoid coalescence having greater ductility than cleavage fracture. Acting in concert within the weld nugget, however, as the load on the tensile specimens increases, the ductility within the particle-poor layers of the nugget is exhausted first. As the particle-poor layers consequently rupture, the load is transferred to the particle-rich layers, and the load-bearing area of the specimen is effectively reduced. The stress on the particle-rich ligaments, therefore, rapidly increases, and the specimen ruptures with little, total elongation.

This model of weld nugget fracture is consistent with the mechanical properties measured for the post-weld heat treatment specimens, especially for the 6101-T6 specimens that were solution heat treated and aged. Here, the load is equally shared between the particle-poor and particle-rich layers of the weld nugget until the bulk yield stress is reached. The lack of ductility of the particle-poor layers constrains the overall elongation of the specimens until their ductility is exhausted at the yield stress. Being brittle in nature, the particle-poor layers yield and rupture at virtually the same load, transferring the load exclusively to the particle-rich ligaments. The sudden rise in stress, therefore, ruptures the particle-rich layers. As a result, the overall elongation is low, 2.4% for the 6101-T6 specimens with post-weld SHT and age, and the yield and ultimate tensile strengths are nearly equivalent in magnitude, 201 and 202 MPa for the yield and ultimate tensile strengths for the 6101-T6 specimens.

As previously discussed, the aluminum extrusions were plated with tin so as to trace the flow of material during the friction stir weld process. The authors should note that the melting temperature of tin is 232°C; therefore, any uncombined-tin not influenced by the temperatures of friction stir weld process would be subject to melting during the post-weld solution heat treatment. Though no evidence of melting was observed within the weld nugget, melted tin was discovered within the material thickness near the back surface of the extrusions, as shown in Figure 9. The measured mechanical properties, however, are presumed to be a direct result of friction stir welding and the subsequent heat treatments and not an ancillary response of the material related to the melting of the tin during solution heat treatment.

Figure 9 - Evidence of melted tin near the back surface of the aluminum extrusions

Conclusions

Tin plated 6061-T6 and 6101-T6 aluminum extrusions were friction stir welded in a corner configuration. The weld nugget exhibited a banded microstructure consisting of alternating layers of material "rich" in secondary phase particles and material "poor" in the particles. Transmission electron microscopy revealed that the particle-rich bands contained a strong presence of tin, while TEM foils taken from the TMAZ, HAZ and base material thickness did not indicate any Sn-containing phases. Since tin is only present on the surface of the extrusions, the particle-rich bands represent surface material that has flowed into the nugget region. The particle-poor bands with no tin, therefore, must be material that originates from within the thickness of the extrusions.

Tensile specimens that received only friction stir welding showed a decrease in all measured mechanical properties when compared to baseline properties for both 6101-T6 and 6061-T6 extrusions. Additionally, all tensile specimens failed on the retreating side of the weld and revealed a fracture morphology dominated by microvoid nucleation and coalescence around iron, silicon-containing secondary phase particles. Post-weld heat treatments, however, increased the yield and tensile strengths, but decreased the elongation. For the 6101-T6 extrusions, post-weld solution heat treatment and age increased the yield and ultimate tensile strengths beyond the baseline values. The elongation, however, was dramatically reduced, and fracture occurred within the weld nugget. Tensile specimens receiving post-weld heat treatments displayed two distinct modes of fracture within the weld nugget, each mode characteristic to the particle-rich or the particle-poor layers of the banded microstructure.

References

1. W.M. Thomas et al., Great Britain Patent Application No. 9125978.8 (December 1991).

2. C. Dawes, W. Thomas, TWI Bulletin 6, November/December 1995, p. 124.

3. M. A. Sutton, B. C. Yang, A. P. Reynolds and J. H. Yan, "Banded microstructure in 2024-T351 and 2524-T351 aluminum friction stir welds - Part II. Mechanical characterization," *Materials Science and Engineering A-Structural Materials Properties Microstructure and Processing*, vol. 364, pp. 66-74, JAN 15. 2004.

4. B.J. Dracup, W.J. Arbegast, in: Proceedings of the 1999 SAE Aerospace Automated Fastening Conference & Exposition, Memphis, TN, October 5–7, 1999.

5. A. von Strombeck, J.F. dos Santos, F. Torster, P. Laureano, M. Kocak, in: Proceedings of the First International Symposium on Friction Stir Welding, Thousand Oaks, CA, USA, June 14–16, 1999.

6. R. S. Mishra and Z. Y. Ma, "Friction stir welding and processing," *Materials Science & Engineering R-Reports,* vol. 50, pp. III-78, AUG 31. 2005.

7. W.J. Arbegast, Z. Jin, A. Beaudoin, T.A. Bieler, B. Radhakrishnan (Eds.), in: Hot Deformation of Aluminum Alloys III, TMS, Warrendale, PA, USA, p. 313 (2003).

8. K. N. Krishnan, "On the formation of onion rings in friction stir welds," *Materials Science and Engineering A-Structural Materials Properties Microstructure and Processing,* vol. 327, pp. 246-251, APR 30. 2002.

9. M. A. Sutton, B. Yang, A. P. Reynolds and R. Taylor, "Microstructural studies of friction stir welds in 2024-T3 aluminum," *Materials Science and Engineering A-Structural Materials Properties Microstructure and Processing,* vol. 323, pp. 160-166, JAN 31. 2002.

10. T. U. Seidel and A. P. Reynolds, "Two-dimensional friction stir welding process model based on fluid mechanics," *Science and Technology of Welding and Joining,* vol. 8, pp. 175-183, JUN. 2003.

11. C. Hamilton, S. Dymek, M. Blicharski, "Comparison Of Mechanical Properties For 6101-T6 Extrusions Welded By Friction Stir Welding And Metal Inert Gas Welding," Archives of Metallurgy, to be published MAR. 2007.

12. ASTM B 317, "Standard Specification for Aluminum-Alloy Extruded Bar, Rod, Tube, Pipe, and Structure Profiles for Electrical Purposes (Bus Conductor)," ASTM International, 100 Barr Harbor Drive, PO Box C700, West Conshohocken, PA 19428-2959, USA, 2006.

13. W.M. Thomas, E.D. Nicholas, S.D. Smith, in: S.K. Das, J.G. Kaufman, T.J. Lienert (Eds.), Aluminum 2001 – Proceedings of the TMS 2001 Aluminum Automotive and Joining Sessions, p. 213, 2001.

14. W. M. Thomas, K. I. Johnson and C. S. Wiesner, "Friction stir welding-recent developments in tool and process technologies," *Advanced Engineering Materials,* vol. 5, pp. 485-490, JUL. 2003.

15. ASTM E 8, "Standard Test Methods for Tension Testing of Metallic Materials," ASTM International, 100 Barr Harbor Drive, PO Box C700, West Conshohocken, PA 19428-2959, USA, 2006.

16. "Aluminum Standards and Data 2000," The Aluminum Association, Inc., 1525 Wilson Boulevard, Suite 600, Arlington, VA 22209, USA, 2000.

17. M. W. Mahoney, C. G. Rhodes, J. G. Flintoff, R. A. Spurling and W. H. Bingel, "Properties of friction-stir-welded 7075 T651 aluminum," *Metallurgical and Materials Transactions A-Physical Metallurgy and Materials Science,* vol. 29, pp. 1955-1964, JUL. 1998.

18. Y. S. Sato, H. Kokawa, K. Ikeda, M. Enomoto, S. Jogan and T. Hashimoto, "Microtexture in the friction-stir weld of an aluminum alloy," *Metallurgical and Materials Transactions A-Physical Metallurgy and Materials Science,* vol. 32, pp. 941-948, APR. 2001.

Aluminum Alloys
for Transportation, Packaging, Aerospace and Other Applications

Alloy
Processing

Aluminum Alloys for Transportation, Packaging, Aeropsace, and Other Applications
Edited by Subodh K. Das, Weimin Yin
TMS (The Minerals, Metals & Materials Society), 2007

APPLICATION OF TIME-TEMPERATURE-STRESS PARAMETERS TO HIGH TEMPERATURE PERFORMANCE OF ALUMINUM ALLOYS

J. Gilbert Kaufman, Zhengdong Long, and Shridas Ningileri
Secat, Inc., 1505 Bull Lea Road, Lexington, KY 40511

Keywords: Aluminum alloys, High temperature properties, parametric analyses

Abstract

For many years, it has been recognized that the creep and stress rupture properties of aluminum alloys m ay be analyzed and extrap olated utiliz ing tim e-temperature param etric relationships. The Larson-Miller Parameter (LMP), based loosely on the rate-process theory, has proven one of the most useful. In this paper we will update the th eory and appl ication of such param eters to aluminum alloys important in marine and transportation application, not only to creep and stress-rupture data but also to other perform ance da ta involving long-tim e e xposure to temperatures above 75 °C (150 °F). Representative data and master curves will be provided, and illustration of the application of param etric relationships to tensile properties and corrosion resistance. The authors will describe tests unde rway to further define the precision and lim its of suc h applications.

Introduction

Because the properties of alum inum alloys are dependent upon both the exposure tem perature and also to the length of time of exposure, the prediction of design values for structures designed to last m any years is a significan t challenge. For relatively shor t-life structures, the need is addressed simply by planning ahead and carrying out a te st plan tha t replic ates the intended service conditions. Th is is feasible for structures whose design life m ight be as m uch as a year or even five years, but it is not very practical for s tructures for which the life expectancy is 10 years or more.

Since the early 1950s, the analyses of long tim e, high temperature data for alum inum alloys, ferrous m etals and superalloys (1-4), has been addressed through the use of ti me-temperature parametric equation s that perm it the conso lidation of data obtained over a variety of temperatures and exposure tim es into a singl e relationship. Once su ch relationships are established based upon the avai lable experim ental data and optimized, it is possible to extrapolate to service conditions substantially be yond the range of the test data them selves. This must always be done c autiously and with awaren ess of the extent of the extrapolation, but it provides a better perspective than simply extrapolating individual strength life curves.

Within the scope of this pape r, the authors will briefly re view the background for and the application of the m ost widely us ed tim e-temperature pa rameter, the Larson-M iller Par ameter (LMP) plus, m ore im portantly, demonstrate that there is value in the application of such parameters to types of data and perf ormance characteristics beyond the creep-rupture data for which they are best known and most widely used.

137

Illustration of the Need for Time-Temperature Parameters

The need for a m ethod of extrapol ating experim ental creep-rupture test data to longer tim es for estimating service life design values may be readily seen by the representative experim ental data shown in Figure 1 for 5454-O. Typically the d ata for each tem perature appear as discrete lines of decreasing rupture stress with increasing time at tem perature. The l ongest experim entally determined rupture lives are in the range of 5000-10,000 hours, whereas service lives are typically at leas t 10 years or approaching 100,0 00 hours. Utili zing th e indiv idual tem perature curves alone, the extrapolations would be at least one full order of magnitude of rupture life.

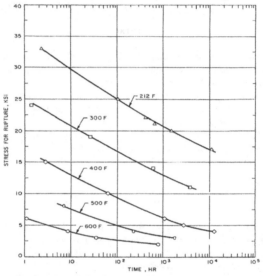

Figure 1 – Creep rupture strengths of aluminum alloy 5454-O

So the advantage of having som e t ype of para metric rela tionship invo lving stress, tim e, and temperature is to provide f or the consolida tion of the individual curves into a single m aster relationship that would enable the ready prediction of the cum ulative effects of both tim e and temperature on s tress. It is p recisely such con solidation that pa rametric analy ses of the type described herein accomplish, and it is the background and broader application of such parameters that are described.

Several time-temperature-stress parameters have been applied with considerable success over the years, especially to stress-rup ture d ata for a va riety of m etals, and to a lesser extent, to creep rates, and total accum ulated creep of various amounts. The three m ost commonly utilized have been the La rson-Miller Param eter (1), the M anson-Haferd Param eter (2), and the Dorn-Sherb y Parameter (3). W hile not necess arily showing great tech nical advantage over th e other two parameters, the Larson-Miller Parameter (LMP) has become the most widely used, for aluminum alloys at least, prim arily because of the ease of its application in iterative analyses to achieve the most useful results. Therefore, in this paper, the authors will focus on the LMP criteria. More detailed information on the applica tion of all th ree param eters, including direct com parisons of their usefulness, is provided in Reference 5.

138

Rate Process Theory and the Larson-Miller Parameter

The early developm ent and application of th e high temperature param etric relationships, including th e Larson-M iller Par ameter, to d ata f or alum inum alloys as well as m any f errous alloys was based upon what was known as the "rate process theory." It was first proposed by Eyring in 1936 (6) and was first applied to metals by Kauzm ann (7) and Dushm an et al (8), expressed as follows:

$$(1) \quad r = A\,e^{-Q(S)/RT}$$

where: r = the rate for the process in question, A = a constant, $Q(S)$ = the activation energy for the process in question, R = the gas constant, T = absolute temperature

In 1963, Manson (9) illustra ted that the LMP and other commonl y used parametric relationships derive from the following form of the rate process equation:

$$(2) \quad P = \frac{(\log t)\,\sigma^Q - \log t_A}{(T - T_A)^R}$$

where: P = a parameter combining the effects of time, temperature, and stress σ = stress, ksi, T = absolute temperature, T_A, Log t_A, Q, and R = constants dependent upon the material

Larson and Miller (1) chose to simplify the rela tionship by pre-selecting values of the four constants as follows:

$Q = 0$; thus $\sigma^Q = 1$
$R = -1.0$
$T_A = -460\ ^{\circ}F$ or $0\ ^{\circ}R$
$\log t_A$ = the constant C in the LMP

Thus the general equation reduces to:

$$(3) \quad P = (\log t + C)\,(T) \text{ or, for the Larson-Miller Parameter,}$$

$$(4) \quad LMP = T(C + \log t)$$

This analys is has the ad vantage th at $\log t_A$ or C is the only constant that m ust be defined by analysis of the data in question, and it is in effect equal to the following at isostress values:

$$(5) \quad C = (LMP/T) - \log t$$

In such a relationship, isostress data (i.e., data for the sam e stress but derived from different time-temperature exposure) plotted as the reciprocal of T vs. log t should define straight lines, and the lines for the various stress values should intersect at a point where $1/T = 0$ and $\log t$ = the value of the unknown constant C.

Larson and Miller took one step fu rther in their original proposal , suggesting that the value of constant C could be taken as 20 for many metallic materials. Othe r authors have suggested that the value of the constant varies from alloy to alloy, and also with such factors as cold work, and thermo-mechanical processing, and phase transitions or other structural modifications.

From a practical standpoint, most applications of the LMP are made by first calculating the value of C that provides the best fit in the parametric plotting of the raw data, and values for aluminum alloys, for example, have been shown to range from about 13 to 27.

Illustrative Application of LMP to Creep Rupture Data

Several interesting facets of the value and limitations of the parametric relationships may be seen from looking at a representative illustration of the application of LMP to the data for 5454-O presented earlier. Alloy 5454 is the highest strength Al-Mg alloy recommended for applications involving high temperatures; with higher magnesium, some susceptibility to stress corrosion cracking may result from long-time high-temperature exposure.

Figure 1 provides a graphical summary of the empirically-determined creep-rupture strengths for 5454-O over the temperature range from room temperature (75 °F or 535 °R) through 600 °F (1060 °R) from Reference 4. The data are plotted as rupture strength as a function of rupture time for each test temperature.

In order to apply the LMP to these data, it is first necessary to calculate the appropriate value of the constant C in the LMP. As noted earlier, while Larson and Miller has judge a generally useful value to be 20, experience has shown that for aluminum alloys, the best approach is to calculate the constant providing the best fit for the available experimental data. This is done by utilizing the graphical presentation in Fig. 1 to identify as many values of stress providing rupture lives at two or more temperatures, referred to as "isostresses."

Table 1 summarizes the isostress calculations to determine the LMP for 5454-O utilizing the data from Fig. 1. Some of the isostress selections involved modest extrapolations of the experimental data, but all appear reasonable based upon the graphical presentation. A range of values of C from 12.7 through 14.9 was observed.

Temperature combination, °F	Stress, ksi	Temp. T_1 (°R)	Time, t_1 (hr)	Temp. T_2 (°R)	Time, t_2 (hr)	Factor C
212-300	20	672	1 560	760	15.5	14.1
	19	672	2 700	760	25.7	14.0
	17	672	11 010	760	81	14.4
300-400	15	760	275	860	2.8	14.7
	11	760	3 800	860	32	14.2
400-500	8	860	250	960	7	12.5
	6	860	1 094	960	35	11.3
	5	860	2 770	960	89	10.9
	4	860	12 850	960	239	12.5
400-600	6	860	1 094	1060	1.1	12.8
	4	860	12 850	1060	8.5	12.7
500-600	6	960	35	1060	1.1	14.4
	4	960	239	1060	8.5	13.0
	3	960	1 312	1060	35	14.9

Table 1 – Isostress calculations to determine Larson-Miller Parameter constant C for 5454-O creep rupture data

140

The scientists carrying out this work (4) elected to use a mean value of 14.3 for C for 5454-O. This was subjective to the extent that this is in the higher end of the range of calculated values of C but reflects a leaning toward values of C associated with the lower stresses and longer creep lives most likely to be involved in extrapolations to actual service conditions, e.g., creep rupture lives in excess of 100,000 hours.

Fig. 2 illustrates the master LMP relationship generated utilizing the value of 14.3 for the constant C. The quality of the fit is generally relatively good, the principal exceptions being several of the data points obtained at 400 °F which fall just slightly below the master curve. This is reflective of the fact that a value of C around 12.5 might better have fit the relationship in the range of 400-500 °F. At any rate the relationship in Fig. 2 appears satisfactory for extrapolations of at least one order of magnitude, perhaps more.

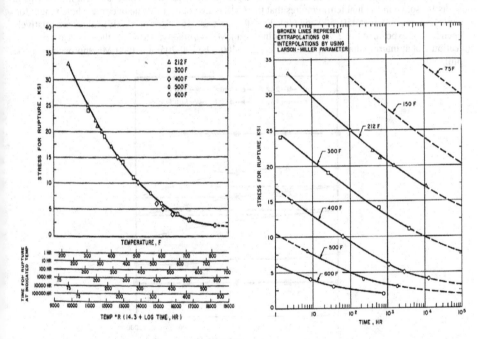

Fig. 2 Master Larson-Miller Parameter (LMP) curve for creep rupture strengths of 5454-O

Fig. 3 – Extensions of creep rupture strength curves for 5454-O by extrapolation using the LMP master curve

Fig. 3 illustrates the extrapolations of the creep stress vs. rupture life curves out to 100,000 hours generated utilizing the LMP with a constant of 14.3.

Applications to Other High Temperature Data for Aluminum Alloys

While the application of the Larson-Miller Parameter and time-temperature parameters to creep data, including rupture life and times to develop specific amounts of creep strain (i.e., 0.1%, 0.2%, 1% etc) is fairly wide spread, little use is made of the parameters in analyzing other types of high temperature data for aluminum alloys.

One obvious example of other high temperature data to which the parameters might be applied is tensile properties at temperatures above room temperature. For aluminum alloys, both the temperature and the time of exposure at temperature affect the resultant values, and the effects of time at temperature are cumulative if the exposure is alternating. The LMP may be of value in extrapolating the effects of exposures longer than those covered by the experimental testing.

Using 5456-H321 as the example, Fig. 4 illustrates the effects of time at different temperatures on tensile strength at temperatures from 212 °F (100 °C) to 600 °F (315 °C). There are clear indications in the plots for some individual temperatures that the LMP may not be of value over the whole range; for example, it appears that at 450, 500 and 600 °F (230, 260, and 315 °C, respectively) values are relatively independent of exposure time, so that blending via LMP is unlikely. However there is also some indication that at intermediate temperatures, say 212-400 °F (100-205 °C), the LMP approach may be helpful.

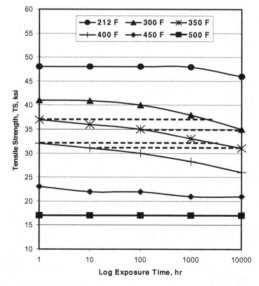

Fig. 4 – Tensile strengths of 5456-H321 at elevated temperatures after various exposure times at the test temperatures

Calculations of the LMP constant C for 5456-H321 tensile strengths leads to quite a wide range of values (~40-65), with an average value of 54. This value leads to the master LMP curve in Fig. 5. The master curve looks remarkably uniform and consistent with most data point, the major exceptions being those for the higher temperatures noted previously. It would appear that for the intermediate temperatures at least, the LMP m ay be a useful tool for long-exposure ex trapolation, but that it must be used with caution, and by careful comparisons with other graphical means of extrapolations.

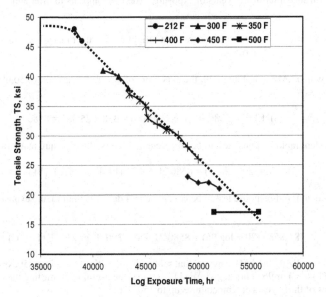

Fig. 5 – Master LMP curve for tensile strength of 5456-H321 at elevated temperatures

Application to Microstructural Changes and Corrosion Performance

While there has been little published on the app   lication of param   eters such as LMP to study microstructural changes from   high tem   perature e xposure or for corrosion perform   ance following extended high tem perature performance, the authors   are presently involved in studies involving both characteristics.

The desirability and usefulness of such an approach is illustrated by m arine exposure experience by the U.S. Navy and Coast Guard in which ships stationed   for years in equatorial environm ents are subjected to endless hours of on-deck tem peratures approaching 150 °F (65 °C). In battle zo nes such as recen t years in the Mideast, high tem   perature exposures are aggravated by   temperature rises in the deck components surrounding gun turrets firing at regular intervals. The net result m ay be the equivalent of 20-30 years of exposure to temperatures averaging 150 °F (65 °C).

Because some aluminum-magnesium alloys, like early versions of 5456-H321, have been widely used in ship superstructures for more that 40 years, such exposures have sometimes resulted in a gradual buildup of beta phase precipitates along th e grain boundaries in such alloys, in turn m aking them susceptible to grain boundary corrosion and exfoliation attack after many years of service. (10)

In order to identify or develop new alum inum all oys and tem pers that are not subject to such grain boundary buildup and resultant corrosion attack, there is a need to esta blish som e short-term test to predict the perform ance after m any years of exposure. The LMP appears to offer a m eans to achieve this.

For exam ple, to determ ine short-term exposures that m ight predict the m icrostructural cond itions after thirty years of exposure at 150°F (65 °C):

- Using LMP with a LMP constant of 20, the expo sure of about 30 years (say 250,000 hr) at 150 ^{0}F (65 ^{0}C) becomes

$$(6) \ \ LMP = 338(20 + \log 250{,}000) = 338 \times 25.383 = 8580$$

- For an equivalent rapid-response test to be complete in 4 hours, the exposure temperature must be:

$$(7) \ \ 8580/(20 + \log 4) = 8580/20.598 = 417 \ ^0K \text{ or } 291 \ ^0F \ (144 \ ^0C)$$

- For an equivalent rapid-response test to be com plete in 4 days (96 hours.), the exposure tem perature must be:

$$(8) \ \ 8580/(20 + \log 96) = 8580/21.976 = 390 \ ^0K \text{ or } 243 \ ^0F \ (117 \ ^0C)$$

These calculations utilizing the LMP suggest that relatively short-time experimental exposures of either 4 hrs at 291 ^{0}F (144 ^{0}C) or 96 hours at 243 ^{0}F (117 ^{0}C) may be useful in predic ting the effect of m arine service exposures of thirty years at temperatures up to 150 ^{0}F (65 ^{0}C)

Tests are under way to explore this approach w ith four Al-Mg all oys, including Mg contents ranging from 3-5%. Som e alloys with Mg contents in the upper end of this range, notably 5456 in the H321 temper, have exhibited susceptibility to exfoliati on and grain boundary corrosion attack as a result of beta phase precipitate buildup along grain boundaries following extended high temperature exposure.

Preliminary results, illustrated by the micrographs in F ig. 6, show that as a result of the 4 and 96 hours exposures at 291 ^{0}F (144 ^{0}C) and 243 ^{0}F (117 ^{0}C), respectively, the m icrostructures of 5456-H116 exhibit greater precipitation and se vere concentration along the grai n boundaries where it would likely lead to grain boundary corrosion attack.

Obviously it will take m any years to prove c onclusively if this appr oach is ac curate and r eliable. Nevertheless, in the short term , i t offers a m eans of e stimating effects that would otherwise be completely unpredictable.

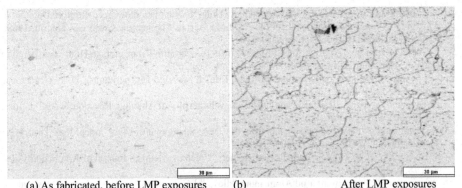
(a) As fabricated, before LMP exposures   (b)   After LMP exposures
Figure 6 – Representative microstructures of 5456-H116 before and after LMP exposures projecting thirty years exposure at 150°F (65 °C)

Conclusions

The usefulness of the use of parametric relationships such as the Larson-Miller Parameter (LMP) for the analysis and extrapolation of high temperature data for aluminum alloys has been described herein, noting the considerable value for creep and stress-rupture strength calculations. An example of the application of LMP to the tensile properties of one alloy (5456-H321) has also been illustrated, indicating its limitations at extreme temperatures (at or above 500 °F, 260 °C) but potential value at intermediate temperatures, say 212-400 °F (~100-205 °C).

A study currently underway is also described in which the LMP is being utilized to estimate the effects of long-term exposure to high temperatures on the microstructure and corrosion resistance of several alloys. The study includes four Al-Mg alloys with 3%-5% Mg, which with prolonged exposure at 150 °F (65 °C) or more, may exhibit grain boundary precipitation and in some cases resultant grain boundary and exfoliation corrosion attack. Short-term exposures have been devised using LMP in order to make a first judgment of the effects of very long time exposures on the corrosion performance of these alloys.

For more detail on the development of time-temperature parameters and their application to aluminum alloys, readers are directed to Reference 5 (to be published in 2007).

References

1. F.R. Larson and James Miller, "A time-Temperature Relationship for Rupture and Creep Stresses," Transactions of the ASME, Vol. 74, ASME, New York, July, 1952, pp 785-771
2. S.S. Manson and A.M. Haferd, "A Linear Time-Temperature Relation for Extrapolation of Creep and Stress-Rupture Data," NACA Technical Note 2890, NACA, March, 1953
3. O.D. Sherby and J.E. Dorn, "Creep Correlations in Alpha Solid Solutions of Aluminum," Transactions of AIME, Vol. 194, AIME, Pittsburgh, 1952.

4. K.O. Bogar dus, R.C. Malcolm , a nd Marshall   Holt, "Extrapolation of Creep-Rupture Data for Aluminum Alloys," presented at the 1968 ASM Ma   terials Engineering Congr ess, Detroit, 1968; ASM Publication D8-100, 1968, pp. 361-390.
5. J. Gilbert Kaufm an, "Applica tion o f Stress-Tim e-Temperature Param eters to High Tem perature Data for Aluminum Alloys, " ASM, Materials Park, OH 2007
6. S.S. Manson, "Design Considerations for Long Life at Elevated Tem peratures," NASA Technical Report NASA TP-1-63, 1963.
7. H. Eyring, "Viscosity, P lasticity, and Diffusion as Exam ples of Absolute Reaction R ates," Journal of Chemical Physics, Vol. 4, p. 283, 1936.
8. W. Kauz mann, "Flow of Solid Metals from th   e Standpoint of Che   mical Rate Theory," Transactions of AIME, Vol. 143, p. 57, 1941.
9. S. Dushm ann, L.W. Dunbar, and H. Huthste iner, "Creep of Metals," Journal of Ap plied Physics, Vol. 18, p. 386, 1944.
10. "Corrosion of Alum inum a nd Alum inum Alloys," J.G. Kaufma   n, Reviewer, ASM Metal   s Handbook, Volume 13B, ASM International, Materials Park, OH, 2005, pp.95-124.

STUDY OF ROLLING BEHAVIOR OF CLOSED-CELL ALUMINUM

FOAM MATERIAL

G. Y. Zu[1], G. C. Yao[2], H.B.Li[3]

[1]Guo-yin. Zu, lecturer of School of Northeastern University,
Liaoning,Shenyang 110004, China, E-mail: zuguoyin@163.com.

[2]Guang-chun.Yao, Professor,

[3]Hong-BIN.Li, Graduate Student

Abstract

Author studied the rolling behavior of closed-cell aluminum foam that was prepared by melt-based route, and analyzed the deforming character and damage manner of aluminum foam at different reduction in pass in this paper. The results showed that the rolling of closed-cell aluminum foam should following the principle of many passes and little deformation, and when the reduction in pass is 0.1mm, the aluminum foam (11mm thickness) consisting of 12wt%Si can realize the total deformation of 1mm. When the reduction in pass is oversize, the damage manner of aluminum foam is typically brittle fracture. The proper reduction advantages in exploiting the excellent energy absorbing characteristic, but when small deforming is carried out bending deformation will occur for aluminum foam during rolling so that the shear damage may occur.

Introduction

Aluminum foam is widely used in aerospace, transportation, construction and military applications, etc., due to its excellent properties such as low density, sound insulation, thermal insulation, electromagnetic wave shielding and non-flammability, etc.[1-4]. Presently, the technologies mainly used to produce Al foam are melt-based route, powder metallurgy process and penetrating casting process, etc.. Melt-based route is always used to produce Al foam with a large size and high mechanical properties[5-7]. Now, Al foam sheet with the length of 1800mm and the width of 800mm can be manufactured in China.

Al-Si alloy is the main material for producing Al foam. Because of high Si content, Al foam is not suitable for rolling. Few researches are focused on the rolling behavior of Al foam. Precise dimension and good shape are very strict in the application field of Al foam. In the present work, the rolling behavior of closed cell Al foam produced by melt-based route was investigated, the fracture mechanisms of Al foam under different deformed conditions were also discussed. A method of improving the rolling ability of closed cell Al foam was put forward.

Experimental

The material used to produce large size Al foam in this work is Al-12%mass Si, the content of TiH_2 as blister is 1.2%. The dimension of the sheets for rolling prepared by saw is 10.5~11.5mm in thickness, 60mm in width and 400mm in length. The rolling tests were carried on Φ180mm rolling mill. Reduction in pass can be controlled precisely, the rolling speed is 0.1~0.3m/s. Due to the poor ductility of Al foam sheet, the rolling ration for each pass reduced for 0.3mm gradually. The rolling test finished when macroscopic fracture was observed. The fracture morphology was collected using the Digital Cameral.

Results and Discussion

Deforming behavior of Al foam under different rolling ration

The effect of the rolling ratio on the rolling behavior of close cell Al foam is listed in Table1. Many transverse cracks are observed when reduction in pass exceeds 0.1mm?. Total reduction could be achieved 1mm after several passes when the reduction in pass is 0.1mm. An improved surface quality and dimension precision are obtained.

It can be found that Al foam is suitable for multi-passes and small deformation. The optimum reduction in pass is 0.1mm. When rolling ratio is too large, the brittle fracture occurs. When reduction in pass was 0.1mm, a deformation ratio of 1mm is achieved after 10 passes. It demonstrates that Al foam can be rolled under small reduction ratio. The width and length of sample change little during rolling, deformation mainly occurs in transverse section, which indicates that closed cell is compacted. When reduction in pass continuously decreases to 0.08mm, the total reduction decreases accordingly. The increasing bending deformation due to decreasing reduction in pass becomes an important failure factor.

Table.1 Rolling experimental results of closed-cell aluminum foam material under different rolling technologies

Initial thickness/mm	Initial width/mm	Initial length/mm	Reduction in pass/mm	Final quality /mm	Length as rolled/mm
11.4	60.8	402	0.3	failure	/
11.2	60.4	403	0.2	failure	/
11.1	60.3	404	0.15	failure	/
11.2	60.2	401	0.1	good	402
11.1	60.2	402	0.1	good	402
11.0	60.2	402	0.1	good	403
10.9	60.2	403	0.1	good	403
10.8	60.2	403	0.1	good	404
10.7	60.2	404	0.1	good	405
10.6	60.2	405	0.1	good	406
10.5	60.2	406	0.1	good	406
10.4	60.2	406	0.1	good	407
10.3	60.2	407	0.1	failure	/
11.3	60.0	402	0.08	good	402
11.22	60.0	402	0.08	good	402
11.14	60.0	402	0.08	good	403
11.06	60.0	403	0.08	good	404
10.98	60.0	404	0.08	good	404
10.90	60.0	404	0.08	good	405
10.82	60.0	405	0.08	failure	/

Figure1 illustrates morphology of the samples after rolling. For better comparison, white mark was used on sample surface. Figure 1(a) illustrates sample state after 5 passes (0.1mm of reduction in pass). The marks on the upper and bottom surface change little after rolling. The marks on the side surface change obviously, the variation in transverse direction is equal to total reduction. Deformation distribution in transverse direction is inhomogeneous. Deformation concentrates on the upper and bottom surfaces, the shape and structure of closed cell change little.

Fig.1　Appearance of aluminum foam under different rolling technologies

(a)-after 5 pass(reduction is 0.1mm)；(b)-after 9 pass(reduction is 0.1mm)；(c)-after 12 pass(reduction is 0.1mm)

　　　Figure 1(b) shows the morphology of the failure sample after 9 passes(0.1mm of reduction in pass). It can be seen that bending deformation is the main failure reason due to the anisotropy of physical property of Al foam results in inhomogeneous deformation between upper and bottom surface.

　　　Figure 1(c) shows the morphology of the sample after 12 passes(0.1mm of reduction in pass). Two cracks run through the transverse section are observed and propagate in a zigzag way (shown in arrow marks). The failure manner demonstrates that the ability of resisting compress of Al foam is better than that of resisting bending. The failure of Al foam shows a homogeneous manner without bending.

Deformation behavior of closed cell Al foam during rolling

　　　The absorbency is one of the special properties of Al foam. Porous and ordered structure of Al foam can excellently adjust the applied load. The deformation models during rolling of Al foam are shown in Figure 2. From Figure 2(a), a ordered closed cell structure is taken on before rolling. When reduction in pass is too large, several cracks run through the transverse section are observed. Deformation mainly occurs on upper and bottom surfaces, the closed cell on those parts are compressed. The structure of closed cell in center part changes little.

　　　The absorbency of Al foam load cannot work when the reduction in pass is large. The ordered structure at local region is destroyed under the transient large load, which breaks down the integrity of adjusting load and finally results in the failure in thickness. The failure morphology of the sample under a large reduction in pass is shown in Figure 3. Many much large size cell are observed at local region(shown by ◇).

　　　The failure model under a reasonable reduction in pass is shown in Figure 2(b). The absorbency of Al foam can display entirely because of no local breakage. The porous structure can homogeneously distribute the stress.

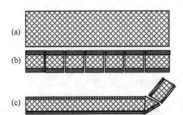

Fig.2 Model of rolling deformation process of closed-cell aluminum foam material

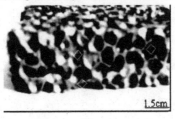

Fig.3 Fracture pattern of aluminum foam at brittle fracture

The bending deformation is generated by rolling in Al foam. The bending parts endure not only the compress stress but also shearing stress caused by bending deformation. With continuous deformation, the cell is failure under large shearing stress, which is shown in Figure 2(c). The crack starts at the bottom surface. Shearing deformation is the main failure factor under a small reduction in pass. The typical failure morphology is shown in Figure 4.

A proposal to improving the rolling ability of Al foam can be put forward: decreasing the Si content, decreasing the cell size, decreasing the reduction in pass with increasing Si content and decreasing rolling speed (optimum rolling speed is 0.1m/s in the present work).

Fig.4 Fracture pattern of aluminum foam at shear fracture

Conclusions

1) The closed cell Al foam with 12%Si content (initial thickness is 11mm) can be rolled under a moderate reduction in pass of 0.1mm, a total reduction of 1mm is achieved after 10 passes.
2) Rolling Al foam is suitable to multi-pass and small deformation. A brittle fracture occurs under a large reduction in pass. The absorbency of Al foam can display entirely under a reasonable reduction in pass.
3) Failure caused by shearing stress takes place at local region under a small reduction in pass.

References

[1]   Zhou Y , Zuo X Q , Sun J L , et al. Cell-structure and mechanical properties of closed-cell aluminum foam[J]. *Transactions of*

Nonferrous Metals Society of China , 2004 , 14（2） : 340-344.

[2]   Jiang B ,Yao G C ,Zhang X M *et al.* Preparation of foam aluminium by powder metallurgy process[J]. *Journal of Northeastern*

University(Natural Science) , 2003 , 24（11） : 1071-1074.

[3]   Ramamurty U , Kumaran M C. Mechanical property extraction through conical indentation of a closed-cell aluminum foam[J].

Acta Materialia , 2004 , 52（1） : 181-189.

[4]   Banhart J. Aluminum foams: On the road to real applications[J]. *MRS Bulletin* , 2003 , 28（4） : 290-295.

[5]   Wang L C , Wang F. Preparation of the open pore aluminum foams using investment casting process[J]. *Acta Metallurgica*

Sinica , 2001 , 14 (1) : 27-32.

[6]   Kitazono K , Nishizawa S , Sato E , et al. Effect of ARB cycle number on cell morphology of closed-cell Al-Si alloy foam[J].

Materials Transactions , 2004 , 45(7) : 2389-2394.

[7]   Song Z l , Nutt S. Energy of compressed aluminum foam[J]. *Advanced Engineering Materials* , 2005 , 7 (1-2): 73-77.

[8]   Kitazono K , Sato E , Kuribayashi K. Novel manufacturing process of closed-cell aluminum foam by accumulative

roll-bonding[J]. *Scripta Materialia* , 2004 , 50 (4) : 495-498.

[9]   Hakamada M , Nomura T , Yamada Y , et al. Compressive deformation behavior at elevated temperatures in a closed-cell

aluminum foam[J]. *Materials Transactions* , 2005 , 46 (7): 1677-1680.

Key words: closed-cell aluminum foam material
rolling
reduction in pass
brittle fracture

Aluminum Alloys for Transportation, Packaging, Aeropsace, and Other Applications
Edited by Subodh K. Das, Weimin Yin
TMS (The Minerals, Metals & Materials Society), 2007

DUAL REFINEMENT OF PRIMARY AND EUTECTIC SILICON IN HYPER-EUTECTIC AL-SI ALLOYS

Mohammad Shamsuzzoha[1] and Frank R. Juretzko[2]

1) School of Mines and Energy Development, Univ. of Alabama, Tuscaloosa, Alabama

2) Dept. of Metallurgical and Materials Engineering, Univ. of Alabama, Tuscaloosa, Alabama

Keywords: Hyper-eutectic Al-Si alloys, refinement, modification, casting

Abstract

Hypereutectic Al-Si alloys with a very fine primary silicon phase and modified eutectic silicon are desirable for their improved mechanical properties. This paper reports on our recent experiments with a hyper-eutectic Al-17wt% Si alloy. The master alloy was cast and cylindrical samples were processed by directional solidification technique at growth rates similar to those found in conventional sand casting conditions. The processed samples have been found to be free of primary silicon. The eutectic silicon in the alloy was found to be very fine and follows a wheat sheaf growth pattern. Individual member in each wheat sheaf assembly of eutectic silicon appears less than 1 micrometer in dimension that is perpendicular to the growth direction. The alloy appears to be a viable candidate for structural applications where improved ductility and machinability is desired. The results of a detailed microstructure analysis by scanning and transmission electron microscopy are discussed.

Introduction

Hyper-eutectic aluminum-silicon alloys posses excellent wear resistance, good electrical and thermal conductivity, a low coefficient of thermal expansion, and a high strength to weight ratio. The alloys also retain most of their mechanical and physical properties at elevated temperatures. These properties make these alloys suitable   se mainly in the automotive industry for the manufacture of engine blocks, pistons, cylinders, brake drums, and hubs [1-7]. These alloys also found use in aircraft engines [8], and in the production of a variety of sand castings and pressure

1

die castings. Further, these alloys exhibit outstanding corrosion resistance and can therefore be used in the manufacture of marine engines and electrical equipment. Table 1 lists the nominal composition of some of the commonly used hypereutectic Al-Si alloys. The major hyper-eutectic alloy is the Reynolds 390 alloy, whose consumption in the North American Continent exceeded 150,000 tons in the year 1992 [9]. In Europe, Japan and the former Soviet Union, alloys with similar composition but different commercial names are widely used for commercial use.

Table 1 Nominal composition of some commercially available hyper eutectic Al-Si alloys

Alloy	Si %	Fe %	Cu %	Mg %	Zn %	Ni %	Mn %	Other
Mercosil	16-19	<1.4	0.15	0.4-0.7	------------	------------	<0.3	------------
390**	17	<1.3	4.5	0.55	<0.1	------------	<0.1	------------
A390***	17	<0.5	4.5	0.55	<0.1	------------	<0.1	
B390**	17	<1.3	4.5	0.55	<1.5	------------	<0.5	------------
392**	19	<1.5	0.6	1	------------	<0.5	------------	------------
393****	22	<1.3	0.9	1	------------	2.25	------------	0.12V
3HA	13-15	0.3-0.4	2.0-2.2	0.4-0.6	------------	2.0-2.5	0.4	0.05Zr, 0.06Ti, 0.04Sr
LM28 (UK)	17-20	0.7	1.3-1.8	0.8-1.5	0.2	0.8-1.5	0.6	0.6Cr, 0.1Pb, 0.1Sn, 0.2Ti, <0.05Co
LM29 (UK)	22-25	0.7	0.8-1.3	0.8-1.3	0.2	0.8-1.3	0.6	0.6Cr, 0.1Pb, 0.1Sn, 0.2Ti, <0.05Co
LM30 (UK)	16-18	1.1	4.0-5.0	0.4-0.7	0.2	0.1	0.3	0.1Pb, 0.1Sn,. 0.2Ti
A-S18 UNG (France)	16.5-19.5	0.75	0.8-1.5	0.8-1.5	0.2	0.8-1.3	0.2	0.1Pb, 0.05Sn, 0.2Ti
A-S25 UNG (France)	23.5-27	0.75	0.8-1.5	0.8-1.5	0.2	0.8-1.3	0.2	0.1Pb, 0.05Sn, 0.2Ti
G-AlSi21CuNiCo (Italy)	20-22	0.9	1.4-1.8	0.4-0.8	0.2	1.4-1.6	0.6-0.8	0.2Ti, 0.1Co
AK21M2.5N2.5 (Russia)	20-22	0.9	2.2-3.0	0.2-0.5	0.2	2.2-2.8	0.2-0.4	0.2-0.4Cr, 0.04Pb, 0.01Sn, 0.1-0.3Ti

* = Balance - aluminum and unlisted impurities
** = Diecasting
*** = Sand casting, Permanent-mold (gravity-die) casting
**** = Sand casting, Permanent-mold (gravity-die) casting, Diecasting

In the microstructure of hyper-eutectic Al-Si alloys two major components, the primary and the eutectic phase, co-exist. The primary phase consists of virtually pure silicon, existing in the form of idiomorphs, while the eutectic structure consists of an aluminum-rich solid solution of silicon and virtually pure silicon, forming the matrix of the microstructure, in which primary silicon idiomorphs are embedded. The idiomorphs of primary silicon can cause considerable tool wear and machining difficulties [10]. The fracture toughness of alloys containing large idiomorphs of primary silicon is rather low. Therefore, these alloys have to be modified to convert the idiomorphs into small and uniformly distributed particles. Much research has been undertaken [11-21] to modify the shape of primary silicon. The most beneficial amongst those appear to be the addition of small amounts of phosphorus [13-16] and sulphur [11, 17]. Phosphorous causes a reduction in the size of the primary silicon by a factor of 4-10, and increases the number of primary silicon crystals and provides for their even distribution in the matrix of the microstructure. The Reynolds 390 alloy is .... .esult of such a phosphorous treatment of a hypereutectic alloy. This treatment, however, brings about a coarsening of the eutectic microstructure [22-23] which is not desirable, because the best mechanical and physical

2

properties can be obtained from those alloys in which both primary as well as eutectic phases are refined. In spite of this draw back, the phosphorous refinement seems to yield alloys with better mechanical properties [24], such as 10-15% higher tensile stress and 100% increase in elongation, than unrefined alloys.

Refinement of eutectic silicon in Al-Si eutectic alloys can be achieved either by impurity modification [25-29], in which trace amounts of sodium or strontium are added into the melt, or by rapid solidification (30-31). The impurity modification, which permits moderate rates of freezing similar to those of sand casting and results in a refined eutectic, is routinely used. Impurity modification of sand cast hypereutectic alloys, although exhibiting a refined eutectic microstructure, still contain a number of primary silicon crystals of various morphologies [32-33]. Impurity modifiers, such as sodium or strontium, also bring about a shift to the coupled region [34-36] of the Al-Si phase diagram towards the silicon phase [37, 38] and therefore shift the eutectic to approximately 14% Si [39]. This phenomenon was considered to be responsible for the formation of a primary modified Al-Si eutectic microstructure during directional solidification of an Al-17% Si-2% Sr alloy [28]. However, the observation of such hypereutectic microstructure is restricted only to directional solidification, and the transformation to commercial sand castings, so far, has not been successful. The 3HA alloy developed by Comalco in Australia and patented in the USA [40] is an example of such an alloy. Besides sodium or strontium, many other elements and combinations of elements have been used for the purpose of reducing the size of the primary silicon and to modify the eutectic phase [41]. Claims of such achievement seem to appear in the literature now on then [42], but none of them have been quantified, nor have been substantiated. Thus, the achievement of hypereutectic alloys with refined primary silicon and a modified eutectic microstructure has remain an elusive goal for numerous solidification and foundry researchers alike, and need for such a dual refinement still exists.

The present research effort is aimed at refining both phases present, namely the primary silicon phase and the Al-Si eutectic phase, by a new concept which takes advantage of a modified behavior of the solid solution.

Concept   finement

3

The new concept involves shifting the equilibrium eutectic point in the Al-Si phase diagram to higher silicon contents which is dramatically different from the conventional methods of refinement, Figure 1a. It can be found that in a binary eutectic reaction, when the melting points of the components differ widely, Figure 1b, the eutectic composition is usually close to the low melting point component and the high melting phase occupies a relatively small proportion of the eutectic by volume [43]. The Al-Si phase diagram, exhibiting a eutectic point at 12.2 at % Si [44] has this characteristic with regard to melting points, with silicon and aluminum phases being the high and low melting point components, respectively. This seems to suggest that if the melting point difference between the melting temperature of aluminum and silicon is reduced, the eutectic composition in the corresponding phase diagram then moves towards the higher silicon content. The melting point of any eutectic component can be lowered by replacing the high melting point component with a solid solution, in which an element or compound is dissolved into lattices of the component concerned. Moreover, if the two eutectic components are solid state solvents for the same solute, the decrease in the melting temperature of each component can be estimated from the amount of solute present. Thus, solute concentration in the components of a binary eutectic diagram can effectively manipulate the melting point of the individual component and can therefore shift the corresponding pseudo-eutectic composition towards the high melting component. The resulting pseudo-binary reaction can then produce a pseudo-eutectic microstructure for the otherwise hypereutectic composition. In this pursuit, a search of binary and ternary phase diagrams containing both the aluminum and silicon as components revealed the existence of a few elements and compounds which could be dissolved into the lattices of silicon and aluminum. These elements and compounds were accordingly employed to obtain hypereutectic Al-Si alloys that contain a microstructure with both refined primary silicon and very refined eutectic.

4

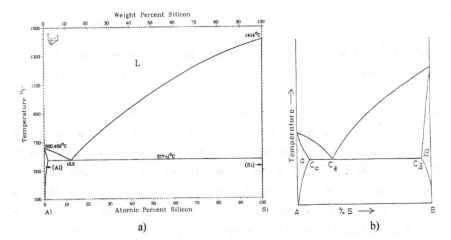

Figure 1 a) aluminum-silicon phase diagram, b) schematic of a phase diagram for faceted/non-faceted eutectic phase diagram.

Experimental Procedures

Hypereutectic alloys in the range of 14 to 17 wt % of silicon were cast in argon atmosphere from aluminum and silicon metal each of 99.999 % purity and selected elements and compounds as discussed earlier. Metallographic examination revealed that the alloys prepared with the addition of the element Ba in the range of 1-4 wt % induces a very fine eutectic microstructure with very little amount of refined primary silicon. The resulting cast billets of nominal composition of Al-17 wt % Si-0.04 wt% Ba were unidirectionally solidified at rates between 100-250 μm s^{-1} and quenched to reveal the growth interface. The temperature gradient at the solid liquid interface varied with growth velocity and was determined by prior calibration of the directional solidification furnace. Longitudinal and transverse specimens taken from near the centre of the grown samples were used for metallography and thin film preparation. Thin foil samples for the TEM (transmission electron microscopy) studies were prepared using the standard preparation procedures of mechanical polishing, dimpling, and ion milling of the specimens to a thickness of less than 100 nm to allow electron beam tran ˙ ion. A Hitachi H-8000 200 keV transmission electron microscope was used. Metallography and determination of chemical composition were performed with a Phillips XL30-SEM (scanning electron microscope).

5

Results

In any typical section of the directionally solidified sample, usual compositional analysis using energy dispersive x-ray spectrometry associated with the SEM, showed an average composition of Al-17 wt% Si, and no trace of Ba in the alloy. Sectioning of the directionally solidified sample and subsequent analysis of the quench interface did not exhibit any primary Si growth, which is the first evidence that the unmitigated growth of the primary phase was successfully inhibited. Figure 2a and 2b show the SEM micrographs of the microstructure from the directionally solidified segment of the sample. The microstructure exhibits only flake morphology for the silicon phase and no idiomorph of primary silicon. The silicon phase in the microstructure follows a wheat sheaf growth pattern, Figure 2b, originating from a point of origin that is typical of that found in unmodified Al-Si eutectic alloys [45]. In the wheat sheaf growth morphology the tendency of the silicon phase is always to diverge from the point of origin. This can also be seen in the high magnification TEM image of Figure 3b.

In the thin film, individual silicon, Figure 3a, has its external surface parallel to internal {111} twins, which is the characteristic of the crystal that is grown by the twin-plane re-entrant (TPRE) mechanism [46, 47], and was observed in the growth of eutectic silicon of unmodified Al-Si eutectic alloys [45, 48]. Furthermore, elaborate TEM studies also revealed that the normal diverging growth tendency of the eutectic silicon as seen in the TEM micrograph of Figure 3b is achieved by small and large angle branching based upon the TPRE mechanisms described for a similar growth habit of eutectic silicon in unmodified Al-Si eutectic alloys [45, 48]. All these information firmly supports that the silicon phase present in the directionally solidified alloy is in fact eutectic silicon of unmodified variety and not of primary silicon. Hence, the alloy grown by the present method is in fact a eutectic alloy but of hypereutectic composition.

The average inter-flake spacing from the SEM micrographs, Figure 2a and b, were measured and found to assume an average value of 2.11 μm for Figure 2a, with a minimum of 1.08 μm. For Figure 2b an average of 2.01 μm with a minimum of 0.83 μm was found. The information thus provided for the alloy suggests a refinement of the eutectic that is considerably finer than those found in unmodified Al-Si eutectic alloys and about still much finer than those found in Na or Sr modified alloys grown at comparable rate of s   ication [49]. Hence, the alloy with such a fine interphase spacing of the eutectic constituents is likely to exhibit marked improvements in physical and mechanical properties over existing hypereutectic Al-17 wt% Si alloys grown with

6

158

a similar growth rate and temperature gradient. The evaluation of the mechanical properties, such as tensile strength and ductility, of this alloy are currently in progress and will be presented as they become available.

a)   b)

Figure 2 SEM micrographs of the transverse section of the deep etched Al-17% Si alloy, directionally solidified , growth velocity $V = 250$ µm/s, temperature gradient $G = 70$ K/cm, a) Silicon phase of flake morphology forming a wheat sheaf configuration, b) radiating clusters of silicon flakes from a point of origin.

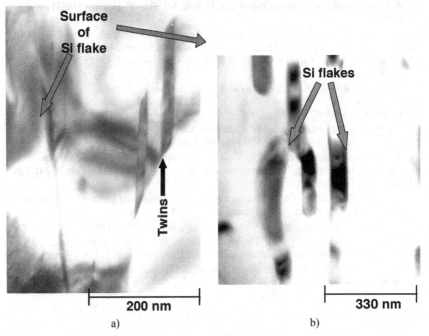

a)   b)

Figure 3 TEM micrographs of Al-17wt% Si alloy, directionally solidified, $V = 1000$ µm/s, temperature gradient $G=70$ K/cm. a) Si-flake with internal twins. b) Diverging Si flakes in the growth of eutectic.

Conclusions

A new concept of refining a hyper-eutectic Al-Si alloy has been presented. It employs the solid solution concept, by which the eutectic point of a binary system can be shifted to the higher melting point component. Using conventional melting and casting production methods a binary hyper-eutectic Al-Si alloy was produced and was subsequently processed by directional solidification at growth rates that were comparable to those found in during conventional sand casting. Characterization of the microstructure by metallography and SEM analysis showed that the microstructure was void of any primary Si idiomorphs and exhibited only eutectic phases as observed. TEM analysis showed that the growth mechanism of the Si phase follows the TPRE mechanism.

Acknowledgements

The authors wish to thank the Central Analytic Facility at The University of Alabama for their support.

References

1. G. Renninger, D. Abendroth and M. Bolien, "Casting Engine Blocks in G K AlSi17Cu4Mg", *SAE International Congress and Exposition*, Detroit, MI, Feb. 28-March 4 (1983) paper no. 830003.
2. P. Hensler, "The New Porsche 9444-Cylinder Aluminum Engine-Application of the 390 alloy Based on the Experience Gathered with Air and water Cooled Porsche Engines," *SAE International Congress and Exposition*, Detroit, MI, Feb. 28-March 4 (1983) paper no. 830004.
3. W. Dahm and R.G. Putler, "Light alloy Engines-Experience Gathered from Research and Development," *SAE International Congress and Exposition*, Detroit, MI, Feb. 28-March 4 (1983) paper no. 830005.
4. E.G. Jacobsen, "General Motors 390 Aluminum Alloy 60° V6 Engine," *SAE International Congress and Exposition*, Detroit, MI, Feb. 28-March 4 (1983) paper no. 830006.
5. H.H. Hofmann, K. Schellmann and E. Wacker, "Aluminum 390 Alloy Engine Blocks: Design and Manufacturing," *SAE International Congress and Exposition*, Detroit, MI, Feb. 28-March 4 (1983) paper no. 830007.
6. J.L Jorstad, "Reynold 390 Engine Technology," *SAE International Congress and Exposition*, Detroit, MI, Feb. 28-March 4 (1983) paper no. 830010.
7. J.L. Jorstad, "Application of 3900 Alloy, an update", *AFS Trans*. 92 (1984) 573-578.
8. R.P. MacCoon and R. Ernst, "The Thunder Aluminum 390 Alloy Aircraft Engine," *SAE International Congress and Exposition*, Detroit, MI, Feb. 28-March 4 (1983) paper no. 83008.
9. J. R. Davis, "Aluminium and Aluminium Alloys", ASM Specialty Handbook, ASM International, Materials Park, Ohio, 1993.

10. J.C. Miller, "Machining High Silicon Aluminum," *11th International Die Casting Congress and Exposition*, Cleveland, OH, June 1-4 (1981) paper No. G-T81-035.

11. A.J. Clegg and A.A. Das, *British Foundrymen*, no. 63 (1970) 56.

12. W. Schneider, W. Reif, and A. Banerji, " Refinement of Silicon Primary Phase in Hypereutectic Al Si Casting Alloys", *Aluminium* (Germany) no. 68, 12 (1992) 1064-1070.

13. C. Mascre, "Modification of High Silicon Aluminum Alloys and the Corresponding Structure," *Foundry Trade Journal*, vol. 94 (1953) 725

14. T. Kawasaki, "Refining of Primary Crystals of Silicon by Adding Phosphorus to Hypereutectic Al-Si Alloys," *Imono*, no. 41, 6 (1969) 434-440.

15. R. Kissling and O. Tichy, *Modern Casting*, vol. 35, 6 (1959) 57.

16. Y. Ikegai, y. Hanyu, A. Hashimoto , and Y. Kitaoka, "76 th Conference of the Japan Institute of Light Metals", Osaka, Japan (1989) 219-220

17. T. Michihiro and S. Yo, "Refinement of Primary Silicon Crystals in a hyper-eutectic Al-20%Si Alloy by Sulphur Additions," *Journal Jap. Inst. Light Metals*, vol. 22 (1972) 564-570.

18. F.L. Arnold and J.S. Priestley, "Hyper-eutectic Aluminum-Silicon Casting Alloys by Phosphorous refinement", *AFS Trans.*, vol. 69 (1961) 129-137

19. H. J. Kim, E. P. Yoon and T. Kobayashi, "Enhanced Modification Process of B390 Casting Alloys by the Ca Content Control ", *Materials Science Forum*, 449-4 (2004) 165-168.

20. J. Chang , I. Moon , C. Choi , " Refinement of Cast Microstructure of Hypereutectic Al-Si Alloys through the addition of Rare Earth Metals", *J. Materials Science*, Vol. 22, No. 20 (1998) 165-168.

21. A. G. C. Gwyer, and H. W. L. Phillips, US Patent 1657389, Jan. 24, 1928.

22. A.P. Bates and D.S. Calvert, "Refinement and Foundry Characteristics of Hyper-eutectic Aluminum-Silicon Alloys," *British Foundryman*, Vol. 59, 3, (1960) 119-133.

23. S. Ghosh and W.J. Mott, "Some Aspects of Refinement of Hyper-eutectic Aluminum-Silicon Alloys," *AFS Trans.*, 72 (1964) 721-732.

24. R.R. Lowery, J.G. Croeni and H. Kato, "Improvement of a Commercial Hyper-eutectic Aluminum-Silicon Master Alloy," U.S. Bureau of Mines, Report no. 6765 (1966)

25. A. Pacz, U.S. Patent 1387900, August 16, 1921.

26. B.M. Thall and B. Chalmars, "Modification in Aluminum-Silicon Alloys," J. Inst. of Metals, 77, (1950), 79-97

27. Shu-Zu Lu and A. Hellawell, "The Mechanism of Silicon Modification in Aluminum-Silicon Alloys: Impurity Induced Twinning," *Metallurgical Trans. A*, 18A, (1987) 1961-1975.

28. D.C. Jenkinson and L.M. Hogan, "The Modification of Aluminum-Silicon with Strontium", *J. Crystal Growth*, 28 (1975) 171-187.

29. M. Shamsuzzoha and L.M. Hogan, "The Crystal Morphology of Fibrous Silicon in Strontium Modified Al-Si Eutectic," *Phil. Mag. A*, 54 (4) (1986) 459-477.

30. S. Gosh and V. Kondic, "Some Aspects of Solidification and Structure of Aluminum-Silicon Eutectic Alloys," *AFS Trans.* 71 (1963)

31. M. Shamsuzzoha and L.M. Hogan, "The Twinned Growth of Silicon in Chill-Modified Al-Si Eutectic Alloys," *J. of Crystal Growth*, 82 (1987) 598-610.

32. G.K. Sigworth, "Observations of the Refinement of Hyper-Eutectic Al-Si Alloys," *AFS Trans.*, 95 (1987) 303-314.

33. N. Tenekedjiev, D. Argo and J.E. Gruzlesi, "Sodium, Strontium and Phosphorous Effects in Hyper-eutectic Al-Si alloys," *AFS Trans.*, 97 (1989) 127-136.

9

34. G. Tamman and A.A. Botschwar, "On the influence of Cooling Velocity on the Structure of Eutectic," *Z. Anorg. Allg. Chem.*, 178 (1928) 425

35. A. Kofler, " Concerning Anomalous Precipitation in Under Cooled Binary Melts and Especially the So-called "Halo Structure'," *Z. Metallk.*, 41 (1950) 221

36. L.M. Hogan, "The Coupled Region Concept in Eutectic Solidification," *J. Aust. Inst. Metals*, 9 (1964) 228

37. R.C. Plumb and J.E. Lewis, "The Modification of Aluminum-Silicon Alloys by Sodium," *J. Inst. of Metals*, 86 (1957-58) 393-400.

38. M.L.V. Gayler, "The Undercooling of Some Aluminum Alloys," *J. Inst. of Metals*, 39 (1927) 157-194.

39. M.M. Haque and V. Kondic, "Influence of Strontium on Solidification of Aluminum-Silicon alloys," *Light Metals: Science and Technology*, Proc. of an Intern. Symposium (1985) 157-170.

40. D.M. Smith, U.S. Patent 4434014, Feb. 28, (1984).

41. E.L. Rooy, "Summary of Technical Information in Hyper-eutectic Al-Si Alloys," *AFS Trans.*, 80 (1972) 421-426.

42. W. Schneider, "Refinement of Primary Silicon of Hypereutectic AlSi Alloys in Direct Chill and Ingot Casting" *Light metals*; Denver, CO, USA, Feb. 21-25, (1993) 815-820.

43. E. Scheil, "Eutectic Crystallization," *Z. Metallk.*, 45 (1954) 298.

44. J.L. Murray and A.J. McAlister, *Bulletin of Alloy Phase Diagram*, 5 No. 1 (1984) 74-84.

45. M Shamsuzzoha, L.M. Hogan and J.T. Berry, "Growth Crystallography of Eutectic Phases in Unmodified Al-Si Casting Alloys," *AFS Trans.*, 69 (1992) 619-629.

46. R.S. Wagner, *Acta Met.* 8 (1960), 57.

47. D.R. Hamilton and R.G. Seidensticker, *J. Appl. Phys.* 31 (1960) 1165.

48. K.F. Kobayashi, P.H. Shingu, and R. Ozaki, Intern. Conf. on Solidification and Casting of Metals, Sheffield, 1979, (Metals Society of London, 1979) 101-105.

49. H. Kaya, E. Cadril, Gunduz "Effect of the temperature gradient growth rate and the interflake spacing on the microhardness in the directionally solidified Al-Si eutectic alloys" *J. Materials Engineering and Performance*, Vol. 12, No.5 (2003) 544-551.

10

COMPOSITIONALLY AND STRUCTURALLY GRADED LAYERS PREPARED BY PLASMA-BASED ION IMPLANTATION ON 2024 ALUMINUM ALLOY

J. X. Liao, Z. Tian, J. Xu, L. Jin

Research Institute of Electronic Science and Technology, University of Electronic Science and Technology of China, ChengDu 610054, P. R. China

Keywords: Graded layer; Plasma-based ion implantation; Diamond-like carbon; Aluminum alloy

Abstract

AlN/Ti/TiN/DLC graded layers have been prepared by pl asma-based ion i mplantation (PBII) or derly with ni trogen / tit anium / nit rogen a nd titanium / carbon on 2024 aluminum alloy. The st ructures of the grad ed layers have been chara cterized by XPS, XRD, TE M, AF M and Ra man spectru m. Th e gra ded layer exhibits gr adually compositional and structural change. PBII respectively with nitrogen / titanium / nitrogen and titanium / carbon can respectivel y form a layer rich in AlN, α-Ti, TiN, and sp^3 bond carbons. Every two adjacent la yers are c onnected with a tra nsition layer with intermediate phase such as T iC, TiCN, CNx, TiAl$_3$, and TiO$_2$. The interm ediate phases are caused by the i mplantation, dif fusion or recoil-i mplantation of nitrogen, ca rbon, oxygen, etc. A proper graded layer has been obtained by optimizing PBII parameters, and corresponds to the most significantly improved hardness and tribological properties.

Introduction

Aluminum (Al) and its al loys have been significantly attracted in a viation and space applications owing to their low specific gravity and high strength-to-weight ratio. But their lo w hardness and low wear resistance has been a great obstacle in the applications. Nitrogen (N) ions implantation into A l and its a lloys to f orm an N- implanted layer rich in AlN (A lN-layer) ha s offe red m ore possibility of widely using them in the applications [1-3]. However, the AlN-layer is thin enough (less th an 0 .3µm) not to satisfy the actu al req uirements. As ex cellent wear-res istant film s, diamond-like carbon (D LC) film s ha ve shown increasing attraction in tribolog ical applications due to their high hardness and elastic m odulus, low frict ion coefficient and lo w wear rate [4-6]. Unfortunately, up to no w ve ry lit tle wor k ha s be en done on the properties of DLC on sof t substrates such as Al alloys because soft substrate materials may not be able to provide adequate support for the hard DLC films, adversely affecting their tribological performance and durability owing to considerable st ructure and property di fference between the soft substrate and the DLC film. Meletis et al [7] h ave used plasma nitriding as a pr ecursor treatment to h arden Ti-6Al-4V substrate, followed by ion-beam deposited DLC film on it, resulti ng in satisfactory improvement in tribological properties. This ha s offe red a good i dea t hat c ompositionally a nd st ructurally graded la yer with DLC f ilm top lay er on soft su bstrate like Al and it s alloys should ma rkedly improve the hardness and tribol ogical properties of sof t subs trates. Our inve stigation ha s confirmed this idea [8] . Also , it ha s bee n found t hat t he improvements in hardness and tribological pr operties are st rongly dependent on th e in terlayer b etween the s ubstrate and th e DLC film and the corresponding processes [8]. It is well known that a proper interlayer can most reduce the s tructure and property di fference be tween the substrate and the DLC film and bring the DLC f ilm unique tribological p roperties in to p lay, and a thick er DL C film can in crease supporting capacity. Thus this paper is to prepare graded layer with thicker DLC film top layer by plasma-based ion im plantation (PBII) on 2024 Al alloy and study its structure, hardness and tribological properties.

Experimental details

The graded layer was prepared us ing DLZ-01 PBII a pparatus described elsewher e [8]. 2024 Al alloy is used as the substrate with the following chemical composition (in wt%): 3.8-4.9Cu, 1.2-1.8Mg, 0.3-0.9Mn, ≤0.5Fe, ≤0.5Si, ≤0.3Zn and balanc e Al. The alloy is first solid-solution-treated at 495 °C for 30min and t hen aging-treated at 175 °C for 10h, resulting in HK $_{0.02N}$ 1.41GPa. The Al alloy sa mples we re finished disk shapes with diameters of 40mm, a nd 5mm high. Prior t o PBII, t he sa mples were m echanically polished t o Ra approxi mately 25nm, ultrasoni cally cl eaned with absolute alcohol the n acetone for 5-10min, an d finally dri ed in air. Sputtered with Ar$^+$ at 2kV bias voltage for 30min to remove the residual contaminations and some surface oxides, all samples were orderly i mplanted wit h T i, deposit ed with T i, im planted with N and simultaneously deposited with T i, a nd de posited wit h C by PBII to for m graded l ayer of AlN/Ti/TiN/DLC. The corresponding parameters were listed in Table I.

Table I The PBII parameters of preparing AlN/Ti/TiN/DLC layer

Processes	Working gas	Working pressure
N-implanted (75kV, 40µs, 80Hz, 3h)	N_2	$3×10^{-2}$Pa
Ti-deposited (75kV, 30µs, 80Hz, 3h)	Ar	$8×10^{-2}$Pa
Ti-deposited and N-implanted (75kV, 30µs, 80Hz, 6h)	Ar and N_2	$8×10^{-2}$Pa
C-deposited (20kV, 30µs, 80Hz, 6h)	C_2H_2: H_2 (8:1)	$6×10^{-2}$Pa

The morphologies of the fil ms we re obtaine d by atom ic force m icroscope (AFM), the DLC f ilm structures were analyzed by Raman spectro scopy, the composition depth distributions were per formed by X-ray photoelectron spectroscopy (XPS) using a differentially pumped 3kV Ar$^+$ ion gun at a raster area of about 4 ×4mm^2, and t he core peaks were fitted with a Ga ussian peak progr am. The details of these analysis me thods were described in [9].

Ball-on-disk dry slidi ng wear e xperiments were per formed in a mbient air at room temperature with a relative hum idity of about 50%. Counter-balls with 4mm in diameter and 25nm i n Ra are commercial alum ina. The nor mal applied l oads of 1- 20N, the reciprocating sliding fre quencies of 5Hz, and the sliding amplitude of 12mm are c hosen. The tribological properties are described with wear life, friction coefficient and wear rate. The details were described in [6].

Results and discussions

XPS composition distributions

Figure 1 illustrates the composition distributions of the as-prepared graded layer. As shown in Fi g.2 (a), the graded la yer is comp osed of a carbon fil m (C-deposited layer or C-layer), a T i-N-deposited la yer (T iN-layer), a T i-deposited layer (T i-layer) and an N-implanted layer (AlN-layer), which corresp ond to the PBII pr ocesses listed in table I. The graded layer is about 1800nm thick. Clearly, the XPS depth profile is quite similar to that previously reported by us [8] except that the C-layer and the TiN-layer significantly thicken, resulting in the graded layer tota l thickness notable increase. The i ncreasing thickness of these two l ayers is related to the fa ct that the correspondi ng time incre ases markedly. This is to reduce th e ef fect of the Al soft s ubstrate on the graded laye r especially on the C-layer. The C-la yer is over 800nm t hick and cont ains almost 100% C atom except a little O in sur face layer due t o exposure in air . This sugge sts t hat the C-layer is pure. The TiN-layer contains a large amount of Ti and N and the amount of Ti is above that of N, which i mplies that the T iN-layer isn't composed of pure TiN. The Ti-layer contains much T i and som e O and N. This i ndicates that the T i-layer i s composed of other phases in addition to pure Ti phase. The AlN-layer contains much N which presents Gaussian-like distribution with near t o 40% atom ic concentration. This means that the AlN-layer is rich in AlN. Also, it is f ound that the C-layer and the TiN-layer are naturally connected with a C-Ti transition layer which shows gradual composition change, as sh own in Figure 1 (b). This is caused by C $^+$ ions implantation into the T iN-layer, which is determ ined by PBII char acteristics [10]. Si milarly, the Ti-layer and the AlN-layer are connected w ith a T i-Al transition layer owing to t he Ti$^+$ ions implantation into AlN-layer, and the Ti-layer and the TiN-layer are connected with a TiN-Ti transition layer owing t o the N$^+$ ions implantation into the Ti-layer, as displayed in Figure 1 (c). The form ation of these tran sition layers is also associated with the diffusion and other factors. It is interesting that the N in the TiN-layer and the AlN-layer and the O in the surface layer of Al alloy can come into the Ti-layer by implantation and diffusion. This is bene ficial to increas e the hardness and supporting capacity of the Ti-layer. It is obvious that this graded laye r not only significantly increases in thickness, but evidently improves in gradient compared to other graded layers reported previously

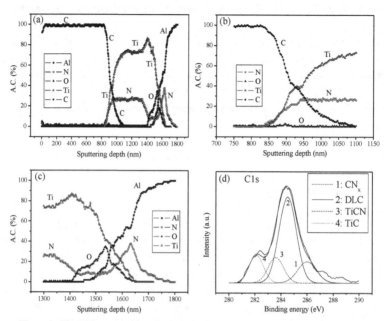

Figure 1: The XPS com position distributions of the as- prepared gradient layer on 2024 aluminum all oy: (a) depth profil e; (b) and (c) the m agnifications of C- TiN and Ti-AlN interfaces in Figure 1 (a), respectively; and (c) the chemical state of C1s at 900nm depth in Figure 1 (b).

[8]. Such a graded laye r on 2024 Al alloy is beneficial to e nhance the connection of each adjacent layer , reduce the di fference bet ween them in compositi on and structure, decrease t he stress between th em, and thus result i n signi ficant i mprovement in tribological properties.

Structures and chemical states

The chemical state of t he core peak of C1s is shown i n Figure1 (d) and those of the core peaks of Ti2p, N1s, O1s and Al2p are shown in Figure 2 (a)-(d) at different depths of the graded layer shown in Figure 1 (a).

As shown in Figure 1 (d), C1s peak can be fitted into four peaks with binding energy (BE) 282.2eV, 283.5e V, 284.6e V and 285.8e V, respectively, corresponding to T iC, TiCN, DLC and CNx [1 1]. At this depth, T i2p peak can be fitted into three pair s of peaks of T i2p3/2 and T i2p1/2 (generall y denoted with T i2p3/2). These T i2p3/2 pe aks show t he BEs of 454.9eV, 455.2e V and 455.6e V and correspond to T iC, TiCN and TiN, respectively. Also, N1s peak c an be fitted into four peaks wit h BE 397e V, 397.5eV, 398.6e V a nd 399.8eV, corresponding to T iN, TiCN, so lid solution N and CNx. I t can be inferred that a long the direction from the C-layer to the T iN-layer C1s gradually shows from DLC to TiC, Ti2p from TiC to TiN, and N1s from TiCN to TiN. Clearly, the C-Ti transition layer exhibits

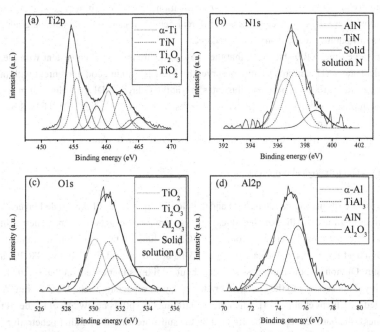

Figure 2: The Chemical states of (a) Ti2p at 1450nm depth of the interface between the TiN-layer and the Ti-layer, and (b) N1s, (c) O1s and (d) Al2p at 1550nm of the interface between the Ti-layer and the AlN-layer in Figure 1 (c).

quite perfect structure transition as well as composition transition, resulting in obvious reduce of structure difference between the DLC and TiN.

Figure 2 (a) displays the chemical state of Ti2p at the interface between the TiN-layer and the Ti-layer. Ti2p can be fitted into four pairs of peaks of Ti2p3/2 and Ti2p1/2. The BEs of Ti2p3/2 are 454.5eV, 455.4eV, 456.9eV, and 458.6eV, respectively, corresponding to α-Ti, TiN, Ti_2O_3 (Ti_2O_3 was produced by Ar+ ion etching TiO_2 during the XPS measurement, and it was unstable and could be easily oxidized into TiO_2) and TiO_2 [12]. At this depth, N1s and O1s show correspondingly consistent chemical states. It can be deducted that along the direction from the TiN layer to the Ti-layer Ti2p gradually shows from TiN to α-Ti. It is obvious that the TiN-Ti transition layer also shows favorable structure transition. Figure 2 (b)-(d) respectively presents the chemical states of N1s, O1s and Al2p at the interface between the Ti-layer and the AlN-layer. N1s can be fitted into peaks with BEs of 397.0eV, 396.5eV, and 398.7eV, respectively, corresponding to TiN, AlN, and solid solution N. O1s can be fitted into four peaks with BEs 529.9eV, 531.3eV, 531.7eV and 532.8eV, respectively. These peaks respectively correspond to TiO_2, Ti_2O_3, Al_2O_3 and solid solution O. Al2p can be fitted into four peaks with BEs 72.6eV, 73.4eV, 74.5eV and 75.5eV, respectively, corresponding to α-Al, $TiAl_3$, AlN, and Al_2O_3. It can be deducted that along the direction from the Ti-layer to the AlN-layer N1s gradually shows from TiN to AlN, Ti2p from α-Ti to $TiAl_3$, O1s from TiO_2 to Al_2O_3, and Al2p from $TiAl_3$ to AlN and finally to α-Al. Apparently, the T i-Al tra nsition layer also exhibi ts good structure transition. From the c hemical stat es at di fferent depths, it can be se en that the graded layer shows quite excellent structure gradient.

At the same time, XRD shows   that the phase structure is   well consistent with the chemical states. The microstructure observed by TEM is also in good agreement with the chemical states. Raman spectrum illustrates th   at the C-layer is DLC f   ilm rich in sp ^{3} carbon atoms, which i s accordant with ot her results reported ear lier [13]. TEM shows that the C-layer are a morphous carbon [8]. The AFM reveals the sur face of the grade d layer smooth and compact owing to the increasing DLC film thickness.

Hardness and tribological properties

Figure 3 e xhibits the nanohardness and tribolog ical properties of the graded layer. As shown in Fi gure 3 (a), the nanohar dness s hows signi ficant change with the penetr ating depths. The hardness i ncreases from 38.6 4GPa at 20nm to th   e maxi mum value o f 42.55GPa a t 80nm, the n decrease s to 31.25G Pa at 150nm, and fina   lly to 9.88GPa at 1600nm. Obviously , the hardne ss shows a bout 30 ti mes incr ease than that of 2024 Al-alloy. This ascribes to the excellent grad   ed layer, especially to the DLC f   ilm. As reported by Chicot et al [1   4], when the penetrating de   pth is below 5% of t   he fil m thickness, the hardness cannot be affected by the substrate, thus the smaller penetrating

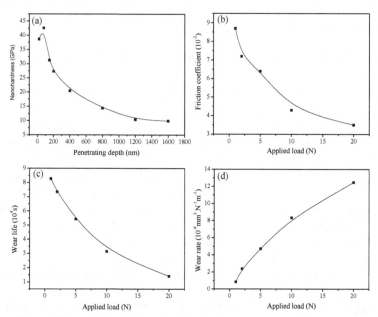

Figure 3: The nanohardness (a) as a function of the pen etrating depth and the tribological properties (b)-(d) as functions of the applied load of the graded layer.

depth corresponds to the more contributions of the films, and vice versa. The hardness of 42.55GPa should be attributed to the intrinsic hardness of the DLC film, which is near to the value of the 180nm-thick DLC f ilm at 10 nm penetrating depth [6]. The hard ness of 38.64GPa at 20nm is lower than 42.55GPa ow ing to the DLC sur face adsorbing oxygen in air. The other hardness should be the mi xed hardness of the gra ded layer a nd the substrate. Such high hardness means high graded layer load-carrying ability.

As displayed in Fi gure 3 (b)-(d), as the app   lied load i ncreases from 1N t o 20N, the friction coefficient decreases from 0.087 to 0.035, the wear life decreases from 82700s to 14200s, and the wear ra te increases from $0.86 \times 10^{-9} mm^3 N^{-1} m^{-1}$ to $12.47 \times 10^{-9} mm^3 N^{-1} m^{-1}$. Such excell ent tribological propert ies also show more significa nt i mprovement tha n those of the thin graded layer reported earlier [8]. Such significant improvement is due to the fact that the i mproved structur e gradie nt and the increased graded layer thi   ckness markedly reduce the str ucture di fference between the DLC fil m and t he substrate and increase the load-carrying ability.

Conclusions

Optimized AlN/Ti/TiN/DLC graded layer has been prepar ed by PBII on 2024 aluminum alloy. The graded layer exhibits gradually compositional and structural change. The DLC film is s mooth and c ompact and rich in sp3 bonds carbon atom s. PBII respectively with nitrogen / titanium / nitrogen and titanium can respectively form a layer rich in AlN, α-Ti and T iN. Every two adjacent l   ayers are connected with a transition layer with intermediate phase such as T iC, T iCN, CN x, T iAl3, and T iO2. The graded layer can show more than 40GPa nanohardness and near to 0.04 friction coefficient at 20N applied load, near to 85000s wear life and 10-9mm3N-1m-1 level wear rate at 1N.

Acknowledgements

The work (Grant No. L080 10301JX05016) was supported by the Y outh Foundat ion o f University of Electroni c Science a nd Technology of Chi na. The aut hors woul d like to thank Harbi n Institute of T echnology for the preparation of DL C fi lms and La nzhou Institute of Chemical Physics of Chinese Academy of Sciences for wear experiments.

References

1. X. L. Fa ng, W. R. Zh i, M. X. Xin and S. Y   ue, "S tructure and W ear Be havior of Nitrogen-Implanted Alum inum Alloys," Journal of Vacuum Scie nce & technology B, 12(2) (1994) 931-934.

2. V. V. Uglov, A. P. Laskovnev, N. N. Cheren da and V. V. Khodasevich, "The Ef fect of Nitrogen Implantation on the Tribological Properties of Composite Aluminum Alloys," Surface & Coatings Technology, 83 (1996) 296-300.

3. L. Guzman, G. Bonini, M. Adami, P. M. Ossi, A. Miote llo, M. Vittori-Antisari, A. M.

Serventi and E. V oltolini, "Mechanical Be havior of Nitrogen-i mplanted Alumi num Alloys," Surface & Coatings Technology, 83 (1996) 284-289.

4. J. Robertson, "Pr operties of Dia mond-like Ca rbon," Surface & Coatings T echnology, 50 (1992) 185-203.

5. A. Grill, "T ribology of Diamond-like Car bon and Related Materials: An Updated Review," Surface & Coatings Technology, 94-95 (1997) 507-513.

6. J. X. Lia o, W. M. Li u, T. Xu, C. R. Y ang, H. W. Che n, C. L. Fu, W. J. Leng, "Structures and tribol ogical pr operties of di amond-like carbon fil ms prepare d by plasma-based ion i mplantation on Si," Surface & Coatings T echnology, 191 (2005) 90-95.

7. E. I. Meletis, A. Erdemir and G. R. Fenske, "Tribological Characteristics of DLC Films and Duplex Plas ma Nitriding / DLC Coating T reatments," Surface & Coat ings Technology, 73 (1995) 39-45.

8. J. X. Liao, L. F. Xia, M. R. Sun, W. M. Liu, T. Xu and Q. J. Xue, "The structur e and tribological properties of gradient layers prepared by plasm a-based i on i mplantation on 2024 Al alloy," Journal of Physics D: Applied. Physics, 37 (2004) 392-399.

9. J. X. Lia o, W. M. Li u, T. Xu, Q. J. Xue, "Carbon nanometer fi lms prepare d by plasma-based ion implantation on single crystalline Si waf er," Applied Su rface Science, 226 (2004) 387-392.

10. J. R. Conard, R. A. Dodd, F. J. Worzala and X. Qiu, "Pla sna Source Ion I mplantation: a New, Cost-effective, Non-line-of-sight Technique for Ion Implantation of Materials," Surface & Coatings Technology, 36 (1988) 927-931.

11. J. F. Moulder, W. F. Stickle and P. E. Sobol, "Handbook of X-Ray Photoelectron Spectroscopy (Eden Prairie, Minnesota: Physical Electronics Inc)," (1995) 210-215.

12. M. W olff, J. W. Schultze and H. H. S trehblow, " Low-energy I mplantation and Sputtering of T iO$_2$ by Nitroge n a nd Ar gon a nd the Electrochemical Reoxidation," Surface and Interface Analysis, 17 (1991): 726-736.

13. J. X. L iao, W. M. L iu, T. Xu, Q. J. Xue, "Char acteristics of carbon fil ms prepared by plasma-based ion implantation," Carbon, 42 (2004) 387-393.

14. D. Chicot and J. Lesa ge, "Absolut e Hardnes s of Fil ms a nd Coatings," Thin Solid Films, 254 (1995) 123-130.

170

Aluminum Alloys for Transportation, Packaging, Aeropsace, and Other Applications
Edited by Subodh K. Das, Weimin Yin
TMS (The Minerals, Metals & Materials Society), 2007

A STUDY OF STABILITY OF FOAM ALUMINUM BY POWDER METALLURGY METHOD

Zhiqiang Guo[1], Guangchun Yao[2], Yihan Liu[3]

[1]Zhiqiang Guo (School of Materials&Metallurgy, Northeastern University);
Shenyang, Liaoning 110004, China
[2]Guangchun Yao Professor, E-mail: G.C.Yao@mail.neu.edu.cn
Shenyang, Liaoning 110004, China
[3]Yihan Liu Associate Professor of Northeastern University

Keywords: powder metallurgy, foam aluminum, viscosity, TiH_2

Abstract

Foam aluminum is a new functional material that has been developed in recent years. It can be used in many fields. In china researches to foam aluminum are on its own way now. In this paper the foaming process of Al-Si-Ca alloy which has high viscosity thus can improve the stability of foam was studied. Major factors influencing the stability of foams were obtained. Adopting appropriate technologic parameters, treating foaming agent (TiH_2) at different temperatures about 2 hours, and then mixing with alloy powders, pressing into matrix, doing foaming experiment, high porosity and stability foam aluminum materials can be obtained by choosing Optimal treating temperature and appropriate parameters

Introduction

Foam aluminum is new functional material that has been developed in recent years. It can be produced by a variety of different methods.[1,2] Powder metallurgy method that can manufacture fairly inexpensive closed cell materials with attractive mechanical properties attached more attention. Fraunhofer-institute for advanced materials of Germany had a deep research on that method, aluminum foam sandwich panels are produced and used to car copings, some semi-manufactured accessories were also produced.[3,4] In china researches on powder metallurgy method were just at its starting period. Further more, major works were concentrated on capabilities of foam aluminum.

In this paper major factors influence the preparation of foam aluminum by powder metallurgy method were studied. Al-Si-Ca alloy which has high viscosity thus can improve the stability of foam was applied

Process of the experiment

Powder Selection

Table I Elements contents of Al-Si-Ca alloy			
Al-Si-Ca alloy	Al	Si	Ca
contents	84.5wt%	12.5wt%	3wt%

Table I shows the elements contents of Al-Si-Ca alloy powders.
The appropriate selection of the raw powders with respect to purity, particle size and distribution, alloying elements, and other powder properties is essential for successful foaming. Commercial air-atomized aluminum powders with d<0.074mm proved to be of sufficient quality.

Experiment Process

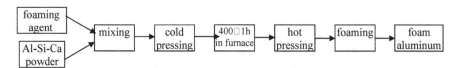

Fig.1: Foam production by powder metallurgy method

Mix the alloy powder with foaming agent (TiH_2) that the amount is 0.6% about 10 hours on tumbling mixer. 100g alloy powders were pressed into a precursor with a diameter of 50mm by cold pressing + hot pressing mode under the pressure of 300MPa. Put the precursor into the experimental set-up for foaming tests, foam aluminum materials were obtained. The precursors were analyzed by SEM. Factors that influence the structures of foam aluminum were studied.

Results and discussion

In this section several fundamental investigations are described to elucidate the foaming processing and also to help find optimized parameter sets for foam production.

Treating temperatures of TiH_2

A related study was carried out by H. W. Gao.[5] She subjected TiH_2 to heat treatments in air at 450~630°C, and chose appropriate treating time, hydrogen is released from pre-treated blowing agents at higher temperatures and in a narrower temperature range than from untreated powder. It also can be concluded from the studies carried out by Biljana Matijasevic[6] 500°C should be the appropriate temperature. So in this paper 500°C heating 3h in air was applied to treat TiH_2

Presently, TiH$_2$ seems to be the best choice for the foaming agent [7]. It can be concluded from the thermo-gravimetric curves for TiH$_2$ powder in argon substantial decomposition must occur at temperature around 400°C or even below. [8] However, massive hydrogen evolution begins only

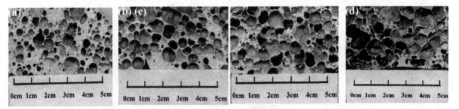

Fig. 2 Evolution of foam aluminum at 680°C

above 450°C and probably ends at the melting point of titanium (1668°C). The cures described in the argon are different from that in the precursor, but it provided a reference.

Fig.2 shows the evolution of foams with the increase of foaming times. The evolution of foam aluminum has gone through three stages:[9] I, formation of aluminum foams: when the temperature is lower than 400□ TiH$_2$ has not decomposed, metallic matrix was still solid state, so it has little expanding. With the increase of temperature, hydrogen starts to release. It is thought that hydrogen gas will first build pore nuclei between surface contacts of former powder particles and at surface contacts to the TiH$_2$ particles. [10] As metallic matrix is further heated, the metal loses its strength and the pores are able to grow as soon as the internal hydrogen pressure have over comes both the resistance of the material and the ambient pressure. The metallic matrix began to expand, as be shown in fig 2a. □, growth of metallic matrix; when the temperature achieved at the melting point of the metallic matrix, metallic matrix began to melt. TiH$_2$ began to decompose rapidly, foams created and grew rapidly until the max expansion as be shown in fig.2b. □, merging and coarsening of aluminum foams; when foams achieved at max expansion, a portion of pore walls became thinner fleetly because of the shortage of resilience, pore walls began to rupture. The gas emitted so much that can not be renewed by the decomposing of TiH$_2$, collapse occurred as be shown in fig.2c and 2d.

Contents of Foaming Agent

Assuming the usual content of 0.6wt% , TiH$_2$ in the foamable precursor material, one find 1cm^3 of the precursor contains 0.016g of TiH$_2$, corresponding to 0.65mg of hydrogen or 0.32mmol H$_2$. At 660°C this quantity of H$_2$ has a volume of 23cm^3. If 75% of the hydrogen in TiH$_2$ is released, a theoretical volume expansion, by a factor of 17 of the originally dense precursor would be expected. As in reality only a little hydrogen is effective in foaming gas-filled pores, whereas the rest is lost during foaming. [5]

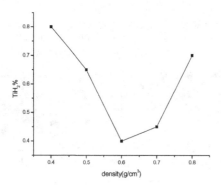

Fig. 3: Density curve with different TiH2 contents at 680°C

Fig. 3 shows the density with different contents of TiH_2 when the precursor got its max expansion. It can be concluded that the more higher of the contents of TiH_2 the lower of the density of precursor. But when the content of TiH_2 is higher than 0.6wt% the density became lower, with the increasing of TiH_2 contents.

Nuclei and growing condition of aluminum foams:

$P_{in} > P_g + P_h + 2\sigma/r$ (1)

In this equation: P_{in}-internal pressure of foams; P_g-atmospheric pressure; P_h-pressure of melt in depth of h; $2\sigma/r$-extra pressure caused by surface tension. [11] It can be concluded that with the increasing of TiH_2 contents, P_{in} became higher. Therefore the max expansion became higher. If foaming agent contents are too high, the surface tension is too low to counteract the expansion pressure P_{in}, collapsing occurred, the density become higher as be shown in fig3. So 0.6wt% would be the appropriate content of foaming agent.

Foaming Temperature

To got high-quality foams, choosing an appropriate temperature was necessary. If the final sample temperature is below the solidus temperature of the alloy, there is not very much more effect than a slight solidstate expansion. If the final temperature lies in the solidus-liquidus interval, foam formation can be observed but is limited to very low expansions. The viscosity of the semimolten material is still quite high at this temperature, and surface oxidation leads to an additional resistance counteracting the internal gas pressure built up by the decomposing foaming agent. Increasingly higher temperatures reduce viscosity and promote gas production so that higher and higher volume expansions can be observed. Besides reducing viscosity, high-furnace temperatures naturally also lead to high heating rates. This can be advantageous for obtaining high volume expansions. If the temperature is much higher than the melting point, the viscosity decreased rapidly, at a high heating rate. Collapsing occurred when the pressure caused by surface tension can not be balanced with P_{in}. [11] The density will become higher. It can be concluded that for high expansion rates, one must ensure a suitable heat flux into the sample up to the maximum expansion by providing a furnace at a sufficiently high temperature.

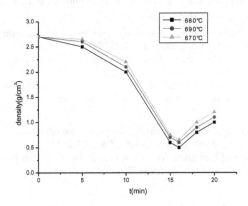

Fig. 4 Density curves of different temperatures

Fig. 4 shows the densities of different temperatures. It can be concluded that 680°C was the appropriate temperature.

Conclusion

Foaming was a complex process. Foaming times, foaming temperature, contents of foaming agent, the treating temperature and time to foaming agent, they jointly influenced the foaming process. In this paper, the relationships of these factors were studied. High quality foams were obtained by choosing appropriate parameters. Form this paper the appropriate conditions were obtained.

1. Foaming agent treating condition: 500°C heating 3h in air.

2. Foaming temperature: 680°C;

3. Foaming time: 16minutes;
4. Foaming agent content: 0.6wt%.

References

[1] Davies G J, Zhen S. "Metallic foams: their production, Properties and Application". *Material Science*, 1983, 18(4), 1899-1903.
[2] Wang Zhutang. "Aluminium foams: production processes, structure and properties, applications and market(3)". *Light Alloy Fabrication Technology*, 1999, 27(12):1-5.
[3] Banhart J. "A design guide metal foams and porous metal structure". *CaseStudies*, 2000, 30(5):217-219.
[4] Baumeister J, Banhart J, "Weber M. Aluminum foams for transport industry". *Materials&Design*, 1997, 18(4):217-220.
[5] Gao Hong-wu; Liu Shi-kui; Zhao Yan-bo; Liu Shun-hua; Li Chang-mao. "Effect of heat oxidation treatment on gas release behavior of TiH$_2$". *The Chinese Journal of Nonferrous Metals*, 2005, 15(3):363-367

[6] B. Matijasevic, J. Banhart. "Improvement of aluminium foam technology by tailoring of blowing agent". *Scripta Materialia*, 2006, 54: 503–508.

[7] F. Baumgärtner, I. Duarte, J. Banhart. "Industrialization of powder compact foaming process". *Advanced engineering materials*, 2000, 2(4):168-174.

[8] F von Zeppelin, M Hirscher, J Banhart. "Desorption of hydrogen from blowing agents used for foaming metals". *Composites Science and Technology*, 2003, 63:2293-2300.

[9] Wei Li, Tang Ji, YAO Guangchun, Zhang Xiaoming, Luo Hongjie. "Evolution of morphology of the foam aluminum in the foaming process by the powder metallurgy method". *Foundry*, 2004, 53 (6) :459-461.

[10] Gao Hongwu. "Preparation of foamed aluminum alloy by powder metallurgy with immersible foaming". *Special Casting & Nonferrous Alloys*, 2005, 25(8):457-459.

[11] Wei Li. "Study on the process of increasing viscosity for producing foam aluminum by the powder metallurgy method". *Foundry*, 2005, 54(3):229-232.

Hydrogen in Aluminum

P.Rozenak, B. Ladna and H.K. Birnbaum

Materials Research Laboratory, University of Illinois, Urbana, Il, USA

E-mail address: rozenak@zahav.net.il

Keywords: Aluminum; Hydrogen; Chemical charging; Defects; Traps.

Abstract

Hydrogen (deuterium) distribution that occurred during electrochem ical and chem ical charging in a high purity single alum inum crystal (grown in [110] direc tion) was studied, using Secondary Ion Mass Spectrom etry (SIMS). Deuter ium distribution was m easured for specim ens that were electrochem ically charged in H $_2SO_4$ solutions or chem ically charged in NaOH solutions, for various charging tim es at room and higher tem peratures. The effect of "aging" related defect form ation in alum inum was de termined. In the hydrogenation of alum inum under high fugacity conditions (such as electrochem ical and chem ical charging), non-steady state diffusion produces concentration-distance prof iles that m ay not be calculated by assum ing simple diffusion behavior. Moreover, the hydrogen- vacancy interactions and m icro structural changes (defect form ations) must be taken into account in the process of the characterizing the state of hydrogen in alum inum. Interstitially hydrogen enters the alum inum lattice poorly. Hydrogen penetrates the alum inum m atrix acco mpanied by a vacancies form ed at the surface during conditions of high fugacity. Alum inum hydroxide and hydrogen interactions form a hydrogen-vacancy com plex at the surface, which diffuses into the volum e and then clusters to form H $_2$ interior bubbles in the alum inum. The SI MS technique was used to characterize hydrogen (deuterium) distributions in chem ically charged alum inum in the order to obtain concentrations-depth profiles. The advantages of this m ethod are that the actual concentrations–depth profiles are obtained and they include m icrostructural changes, such as defects (vacancies, voids, bubbles, micro-cracks, dislocations and surface oxides) formed during the electrochemical and chemical reactions in aluminum with aqueous solutions.

1.Introduction

The goal of the present work was to study the effect of the presence of defects (voids, bubbles, hydroxide), generated by hydrogen-vacancies-alum inum interactions and the presence of m icro-cracks and dislocations, on transient hydrogen (deu terium) diffusion in single crystal chem ically charged alum inum. As has been am ply demonstrated, the calculation of H diffusion in Al, both experimentally [1-5] and in theory [6,7], shows large scatter. Reported values of D $_0$ range from $1.9x10^{-5}$ to $1.75x10^{-8}$ m^{2}s^{-1} and the diffusion enthalpies vary between 16.2 and 140 kJm ol^{-1} [4]. Vacancies, voids, bubbles and oxide hydrides, m ay be generated by the chem ical reactions with the high fugacity presence of hydrogen (deuterium) atoms in the aluminum, while the kinetics of the hydrogen diffusion is com plicated by the hydrogen-vacancy site interactions [7], which are dependent on time and electrochemical conditions.

In many electrolytic charging experim ents, large hydrogen concentrations, approaching atom ic ratio (H/metal) of θ=1 are obtained and sim ple theory is clearly not appropriate [8-10]. In fact, very high pressure techniques have also been used to produce values of θ approaching 1 in gas charging, as was shown by Baranowsky et al. [11]. High hydrogen fugacity on the surface of aluminum sam ples during electrochem ical and chem ical charging [12], introduced high concentrations of hydrogen (~ 1000 appm), and t ogether with X-ray diffraction m easurements, show a sm all decrease or zero change in lattice parameters. This finding m ay be the reason for vacancies, voids or bubbles form ation in Al-H system [13,14]. In the hydrogenation of aluminum, under high fugacity conditions (such as electrochemical and chemical charging), non-steady state diffusion produces concentration-di stance profiles that cannot be calculated by assuming sim ple diffusion behavior. Moreover, the hydrogen-vacancies interactions-related micro structural changes (defect form ation) must be taken into account when characterizing the state of hydrogen in aluminum.

In this study, we use Secondary Ion Mass Sp ectrometry (SIMS) m easurements of deuterium, hydrogen, oxygen and alum inum profiles, formed during electrochemical and chemical charging to measure related defects trapped in the microstructure of a single crystal of aluminum.

2. Experimental procedure

Pure aluminum (99.999%) specimens were used in this study. For som e experiments, specimens were cut from a single crystal i ngot grown in direction [110] in vacuum by a Bridgem an method. The only other detectable elem ents were Si, Fe and Cu, with corresponding concentrations of 1, 2.2 and 1 ppm . The ingot was cut in to pieces, by spark erosion and was followed by careful polishing, to m inimize the dam age to the surface. Care was taken to avoid the introduction of hydrogen during the polishing; thus contact with wate r was avoided at this stage. Specim ens used for secondary ion mass spectrometry were 2mm thick, flat with mirror-like surfaces.

Analyses of the Deuterium (D) distribution were carried out using a SIMS method with a camera IMS 3f ion m icroprobe with the sa mple cooled to 140K. A 17 KeV Cs $^+$ primary beam was used for depth profiling of the sam ple beneath the surface. For the Cs prim ary ion, the depth sensitivity of the analysis was about 5 nm . The prim ary ion beam was rastered over an area of 250x250 µm while the secondary ions were analyzed from the central 10x10 µm, in order to elim inate spurious ions from the crater edges and to obtain the best lateral resolution. The data were normalized using the $^-Al^{27}$ by counts to m inimized instrum ental eff ects. The depth scale was determ ined by measuring the sputtered craters using a stylus profilometer. The depth sensitivity m easurements were in the range of 10 $^{-7}$cm. Deuterium was introduced into the SIMS sam ples by chem ical charging, using 0.1 N NaOD (D $_2$O) solutions and by electrochem ical charging, using 50 m A/cm^2 current densities in 1N D$_2$SO$_4$ (D$_2$O) solution containing 0.25 g/l of NaAsO$_2$ as a D recombination "poison" at room temperature and in higher tem peratures for various tim e periods. After chem ical charging, the samples were kept at 77K to avoid D loss and redistribution before being transferred into the SIMS instrum ent for analysis. Th e concentrations of hydrogen introduced during electrochemical and chemical charging were monitored by gas extraction analysis.

Al single crystal alum inum was used in the measurements according to the Laue m ethod. Mo white radiation was used. In this technique, each di ffracted beam is selected out of the incident beam of white radiation by the d-spacing (distance between adjacent planes) and θ (Bragg angle) value of the crystal planes that produce the reflection. Back reflection techniques were used. The single crystal was free from defects and was pa rallel to the crystallographic [110] norm al (Fig.1). Other experim ents utilized polycrystalline sheet s having a thickness of 0.4-0.5m m. All of the polycrystalline m aterials were well annealed in a vacuum of 10 $^{-5}$ Pa. Lattice param eter measurements were m ade im mediately after charging [12], using step scans of the X-ray peaks obtained with Cu K_α radiation. The lattice param eters were calculated from lines at various Bragg angles, θ, to increase the precision of measurement of $\Delta a/a_0$; which was estimated to be 10^{-4}.

Length change measurements were made during hydrogen charging [12], using a linear variable differential transform er (LVDT). Lengt h change m easurements were m ade along the length of the sheet specim ens, while the hydroge n entered through the lateral surface only. The estimated precision of the $\Delta l/l_0$ measurements is $\sim 5 \times 10^{-7}$ (for the LVDT measurements).

Fig.1. Laue back reflections obtained from single crystal of aluminum.

3. Results and discussion

In all our experiments deuterium was introduced into the aluminum from aqueous solutions [12], using electrochem ical and chem ical charging m ethods. Electrochem ical reactions differ from chemical reactions, in that they involve, apart from the chemical reagents, an electric reagent (the negative electrons) which acts at the interface be tween a m etal (or another phase with m etallic conduction) and solution of the electrolytes. So lubility of alum inum and its oxides can be controlled by [15]:

$$(2) \qquad 2Al^{+++} + 3H_2O = Al_2O_3 + 6H^+$$

$$(3) \qquad Al_2O_3 + H_2O = 2AlO_2^- + 2H^+$$

$$(4) \qquad Al = Al^{+++} + 3e^-$$

$$(5) \qquad Al + 2H_2O = AlO_2^- + 4H^+ + 3e^-$$

Aluminum decomposes water with the addition of sufficiently acidic solutions, resulting in the evolution of hydrogen, dissolving as trivalent Al^{+++} ions and leaving the electrons on the m etal. In the presence of sufficiently alkaline solutions (Na OH), aluminum also decomposes water with the evolution of hydrogen, dissolving as alum inate ions AlO_2^- and leaving the electrons on the electrode. Alum inum oxide or alum ina, Al_2O_3 occur in various form s on m etal alum inum. The physical and chem ical properties of alum ina depe nd, to a large extent, on the tem perature, time, chemical and others conditions during its prepara tion. When alkali is added to an acid solution, an aluminate precipitate is obtained, which is th e hydroxide gel, corresponding practically to the composition Al(OH)₃ and am photeric in nature. However, this alum inum hydroxide gel is not stable and it crystallizes in the course of tim e, to give first, the m onohydrate γ-Al₂O₃×H₂O or boehmite, crystallizing in the rhom bohedral system, then the trihydrate Al ₂O₃×3H₂O or bayerite crystallizing in the m onoclinic system. This de velopment of alum inum hydroxide is known as "aging". In acid or alkaline solution, the alum inum will be attacked as soon as the oxide film is eliminated. This dissolution is slower in acid than in alkaline solutions [15]. On the m etal-oxide interface in alum inum, atom ic hydrogen com bines to form m olecular hydrogen in the porous hydroxide layer [4]. This process effectively coherences the interface and, as a re sult, surface film blistering (surface bubbles) is initiated. Surface bubbles are pressurized by the hydrogen gas and extra m aterial for the surface bubble wall is pr ovided by the low density porous hydrous oxide. The compact barrier layer cannot be penetrated by the gas molecules and hence, a high pressure of gas can develop within the surface bubbles (bliste rs). However, m any works have been published dealing with aluminum hydride formations [16-23].

In the present study, the deuterium distri bution through the thickness of the specim en during charging at room and various tem peratures and ag ing at various tem peratures was characterized by using the SIMS m ethod. A num ber of D de pth profiles were obtained for each sam ple. Typical results are shown in Fig.2, for distribution of ⁻O¹⁶, ⁻Al²⁷, ⁻H¹ and ⁻D² in a single crystal of Al electrochemically charged for 15 min. at room temperature. Ions of oxygen ⁻O¹⁶ (upper curve) showed m aximum intensity on the surface of the sample and then decreased exponentially to a depth of 0.4 μm. Only sm all changes in the oxygen concen trations were observed in the deeper layers. Ions of ⁻Al²⁷ exhibited m aximum values on the sample surface and then decreased exponentially to a depth of 0.4 μm. Ions of hydrogen ⁻H¹ could be found on the alum inum sample during the SIMS m easurements. This hydrogen exhibited maximum values at the surface of the sam ple and then decreased linearly to a depth of the 0.25 μm. Hydrogen can be form ed from moisture that is in the air [24], and diffused into the specim en during the ion irradiation of the high purity alum inum [25]. Electrochemical charging of the single crystal of alum inum for 15min. at room temperature exhibited the form ation of a high concentration of the deuterium on the surface of the sam ple, which decreas ed linearly to a depth of about 0.55 μm. The craters cross section profile was about 0.55 μm deep and 200 μm wide. On the surface of the alum inum, layers of alumina hydroxide Al(OH)₃ and Al(OD)₃, about 0.1 μm thick were formed.

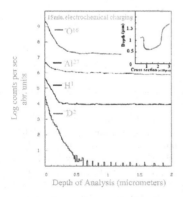 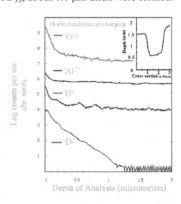

Fig.2. (left side) SIMS profiles of ⁻O¹⁶, ⁻Al²⁷, ⁻H¹ and ⁻D² versus depth in a single crystal of

aluminum electrochemically charged for 15min at room temperature and a crater profile was made by profilometer.

Fig.3. (right side) SIMS profiles of $^-O^{16}$, $^-Al^{27}$,$^-H^1$and $^-D^2$ versus depth in a single crystal of aluminum electrochemically charged for 1h at room temperature and a crater profile was m ade by profilometer.

Typical results are shown in Fig.3, for distribution of $^-O^{16}$, $^-Al^{27}$,$^-H^1$ and $^-D^2$ in single crystal of Al electrochemically charged for 1h at room temperature. Ions of oxygen $^-O^{16}$ (upper curve) show a m aximum value on the surface, decr easing linearly to a depth of 0.1 μm, with negative high slope value. Then the slope value decreases slightly to a depth of about 0.7 μm under the specimen's surface. From that point, the values of oxygen concentration maintain the same level in the deeper layers. Ions of $^-Al^{27}$ exhibited m aximum values on the sam ple surface and then decrease exponentially to a depth of 0.08 μm. Only sm all changes in the alum inum concentration are observed in the deeper layers. Ions of hydrogen $^-H^1$ can be found in the alum inum sam ple during the SIMS m easurements. The hydrogen exhi bits m aximum value at the surface of the sample and then decreases with variati on in the intensity to a deep of about 1μm. From that point, the values of hydrogen concentration m aintain th e sam e level. Electrochem ical charging of a single crystal of alum inum for 1 h at room temperature causes the form ation of a high concentration of deuterium on the surface decreasing linearly to a depth of the 0.08 μm, and then decrease to lower deuterium penetra tion zones, to a depth of about 1.25 μm. The craters cross section profile was about 1.25 μm deep and 200 μm wide.

The results are shown in Fig.4, for distribution of $^-O^{16}$, $^-Al^{27}$,$^-H^1$ and $^-D^2$ in single crystal of Al electrochemically charged for 2h at room tem perature. Ions of oxygen $^-O^{16}$ (upper curve) show a maximum value on the surface, decr easing linearly to a depth of 0.1 μm with negative high slope value. The slope value then decreases slightly to a depth of about 0.6 μm under the specim en's surface. From that point, the values of oxygen concen tration maintain at the same level. Ions of $^-Al^{27}$ exhibit m aximum values on the sam ple surface and then decrease exponentially to a depth of 0.08 μm. Only sm all changes in the alum inum con centration are observed in the deeper layers. Hydrogen ion $^-H^1$ exhibited a m aximum value at the sam ple surface of alum inum and then decreased linearly to a depth of the 0.08 μm, and then decreases with variation in the intensity to a deep of about 1 μm. From that point, the values of hydr ogen concentration m aintain the sam e level. Electrochem ical charging of a single crys tal of alum inum for 2 h at room tem perature exhibits the form ation of high concentrations of deuterium on the surf ace of the sam ple and decrease linearly to a depth of 0.08 μm. They then decrease in lower deuterium penetration zones, to a depth of about 1.4 μm. The craters cross section profile is about 1.4 μm deep and 200 μm wide.

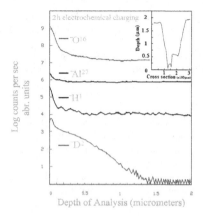 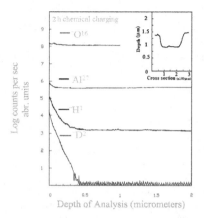

Fig.4. (left side) SIMS profiles of ⁻O¹⁶, ⁻Al²⁷, ⁻H¹ and ⁻D² versus depth in a single crystal of aluminum electrochemically charged for 2h at r oom temperature and a crater profile m ade by profilometer.

Fig.5 (right side) SIMS profiles of ⁻O¹⁶, ⁻Al²⁷, ⁻H¹ and ⁻D² versus depth in a single crystal of aluminum chemically charged for 1h in NaOD solu tion at room temperature and a crater profile done by profilometer.

Typical results are shown in Fig.5, for distribution of ⁻O¹⁶, ⁻Al²⁷, ⁻H¹ and ⁻D² in a single crystal of Al chemically charged for 2h in NaOD solu tion at room temperature. Ions of oxygen ⁻O¹⁶ (upper curve) show small variation in the value on the surface. The level of the oxygen ions concentration is lower in som e levels and closer to the   surface, in com parison to the oxygen concentration obtained in the surface layers in el ectrochemically charged specimens (Figs.2-4). From that point, the values of oxygen concentration m aintain a cons tant level in the deeper layers. Ions of ⁻Al²⁷ exhibited maximum intensity on the sample's surface and then decrease exponentially to a depth of 0.08 µm. Only small changes in the aluminum ion concentration are observed in the deeper layers. Ions of hydrogen ⁻H¹ can be found in the alum inum sam ple during the SIMS m easurements. Hydrogen exhibits a m aximum value at the sam ple surface and then decreases exponentially to a depth of about 0.5 µm. From that point, the values of hydr ogen concentration m aintain the sam e level. Electrochem ical charging of a single crys tal of alum inum for 2 h at room   tem perature exhibits the form ation of high concentrations of the deuterium on the surface, decreasing linearly to a depth of 0.5 µm. The craters cross section profile was about 0.5 µm deep and 200 µm wide.

The form ation of m icro-cracks in high pur ity aluminum during electrochem ical charging by hydrogen was studied [26]. The experiments reveal that in aluminum samples a wide distribution of hydrogen bubbles on the surface (blisters) and   under the surface into the volum e, were produced during electrochemical charging. This phenomenon can lead to the formation of micro-cracks in the absence of externally applied st ress. Exam ination of electrochem ically charged samples by transm ission electron m icroscopy (TEM), showed m icro-cracks with a typically ductile m ode of fracture. Highly plastically def ormed volum es ahead of the crack tip, indicated the appearance of extrem ely high dislocation   density zones. It is know that hydrogen can enhance local plasticity in alum inum [27]. In the work of Bong et al.   [28], in which hydrogen enhanced fractures of age-hardened 7075 and 7075 Al-Zn-Mg alloys were studied, the hydrogen induced fractures were sim ilar to those observed in vacuum , except that they occurred at lower stresses, due to the hydrogen enhanced dislocation mobility.

The depth of these defects was controlle d by electrochem ical charging and depended on the charging tim e. Hydrogen only entered the alum inum lattice interstitially in a weak m anner [13,14]. Either a sm all contraction or zero change in the lattice param eter resulted, when high hydrogen concentrations were introduced into the aluminum matrix. In Al-H solid solution, the vacancy concentrations produced becom e many orders of m agnitude larger than those generated in the H-free lattice at room temperatures below 350K [29]. This is consistent with the formation of hydrogen-vacancy com plexes at the surface. The diffusivity of this com plex at 300K for the plasma charged sample is estimated to be $D_{H-V}=3.9 \times 10^{-15} m^2 s^{-1}$[30].

In the work of W atson et al. [31], on the study of cathodic chargi ng effects on the m echanical properties of alum inum, it was found that the cathodic charging produced a severely hardened surface region. The hardness in this region b ecame quickly saturated, and further charging increased the depth of this region. After electrochemical charging for 24h, the hydrogen determination was found to be at a concen tration of 1000 appm in alum inum at 300K [12]. The concentrations of hydrogen introduced during el ectrochemical and chem ical charging were monitored by gas extraction analysis. It was found [12], that during hydrogenation of Al, an Al(OH)₃ formed during electrochem ical and chem ical charging and other hydrides (AlH, AlH ₃) during hydrogen ion implantation were formed [32]. In the work of Buckley and Birnbaum [30], the surface layer was found to be Al(OH ₃) and on the heating for the gas chrom atography measurements the hydroxide decom posed to Al ₂O₃ and H ₂ resulting in an erroneously high measurements concentration of hydrogen. Al(OH ₃) layer form ed on the alum inum surface was removed by m echanically wiping the surface i mmediately after electrochem ical charging. Moreover, they suggested that in plasm a charged sam ples do not form an Al(OH₃). In this work we were characterized actual distribution of deuterium (hydrogen) in electrochem ically and chemically charged alum inum. This am ount of hydrogen can enhance vacancies, voids and bubbles form ation at room tem perature. The effect of charging tim e is to increase the depth of deuterium distribution under the surface of the alum inum [14]. The deuterium depth is controlled by the electrochem ical charging conditions and th e period of charging. The effective diffusivity of the deuterium (hydrogen) depends on the in terstitial condition, and traps (vacancies, voids, bubbles and dislocations), defects distribution and the hydroxide bubble formation on the surface of the charged aluminum.

The SIMS technique was used to characterize hydrogen (deuterium) distributions in chemically charged alum inum in order to obtain the concentra tions-depth profiles, in ways sim ilar to other techniques such as desorption [1-6, 33-37], permeation [38,39] and other chem ical methods [5] for obtaining diffusion data. The advantage of the SI MS technique lies in the fact that we obtain actual concentrations–depth profiles, including microstructural changes such as defects (vacancies, voids, bubbles, m icro-cracks, disloca tions, surface oxides) that form during the electrochemical and chemical reactions.

4. Conclusions

1. In the hydrogenation of alum inum under high fugacity conditions (such as electrochem ical and chemical charging) non-steady state diffusion produces concentration-distance profiles that may not be calculated by assuming simple diffusion behavior. Moreover, the hydrogen-vacancies interactions related m icro structural changes (def ects form ation) must be taken into account in the process of characterizing the state of hydrogen in aluminum.

2. Hydrogen entered the alum inum lattice inters titially in a weak m anner. Hydrogen enters an aluminum m atrix, accom panied by a vacancies fo rmed at the surface, during high fugacity conditions. Interaction occurs between alum inum hydroxide and hydrogen, form ing a hydrogen-vacancy com plex at the surface, which diffuses in to the volum e and then clusters to form H_2 interior bubbles in the aluminum.

3. The SIMS technique was used to char acterize hydrogen (deuterium) distribution in chemically charged aluminum to obtain the concentrations-depth profiles. The advantages of this method lie in obtaining actual concentrations–depth profiles, including m icrostructural changes, such as defects (vacancies, voids, bubbles, m icro-cracks, dislocations, surface oxides) that form during the electrochemical and chemical reactions in aluminum with aqueous solutions.

Acknowedgments

This work was supported by the Departm ent of Energy, US and thanks the Materials Research Laboratory Center for Microanalysis, University of Illinois for use of their facilities.

5. References

1. B.R. Mclellan, Scripta Metallurgica,17 (1987) 1237.
2. E.Hashimoto, T. Kino, J. Phys. F: Met. Phys., 13 (1983) 1157.
3. T.Ishikava, R.B.McLellan, Acta Metall., 34 (6) (1986) 1091.
4. G.A.Young Jr., J.R.Scully, Acta Mater. 46 (18) (1998) 6337.
5. M.J. Danielson, Corrosion Science, 44 (2002) 829.
6. D. Zang, R.B. Mclellan, Acta Mater. 49 (2001) 377.
7. D Zang, P.Maroevic, R.B.Mclellan, J. of Physics and Chemistry of Solids,60 (1999) 1649.
8. Y.Oren, E.Elish, A.Tamir, Z.Gavra, J. of Alloys and Compounds, 235 (1996) 30.
9. B.Baranowski, Z.Szklaraska-Smialowska, M. Smialowski, Bull.Acad. Polon. Sci. Ser. Chim. Geol., 6 (1958) 179.
10. K.Farrel, M.B.Lewis, Scripta Metall., 15 (1981) 661.
11. B.Baranowski, S.Majchrzak, T.B.Flanagan, J. Sci. Engng. 6 (1998) 141.
12. H.K.Birnbaum, C. Buckley, F. Zeides, E. Sirois, P.Rozenak, S. Spooner, J.S.Lin, J. of Alloys and Compounds, 253-254 (1997) 260.
13. C. Buckley, H.K.Birnbaum, S.J.Lin, S.Spooner, D.Bellaman, P.Starton, T.J. Udovic, E. Hollar, J. Appl. Cryst. 34 (2001) 119.
14. P.Rozenak, E.Sirois, B. Ladna, H.K.Birnbaum, S.Spooner, J. of Alloys and Compounds, 387 (2005) 201.
15. M.Poubaix, Atlas of Electrochemical Equilibria in Aqueous Solutions, Pergamon Press, 1966.
16. P. Breisacher, Siegel B., J. Am. Chem. Soc., 85 (1983) 1705.
17. A.E. Finholt, C.A. Bond, I.H. Schesinger, J. Am. Chem. Soc., 69 (1947) 1199.
18. M. Appel and Frankel J.P., The Journal of Chemical Physics, 42,11 (1965) 3984.
19. P.J. Herley, O. Christofferson, J.A. Todd, J. of Solid State Chemistry, 35 (1980) 391.

20.P.J. Herley, O. Christofferson, J. Phys. Chemistry, 85 (1981) 1887.
21. P.J. Herley, O. Christofferson. and Irwin R., J. Phys. Chem., 85 (1981) 1874.
22. P.J. Herley, O. Christofferson, J. Phys. Chem., 85 (1981) 1881.
23. P.J. Herley, J.A. Todd, J. of Mater. Science Letters, 1 (1982) 163.
24. M.R. Louthan, A.H. Dexter, Metal. Trans. 6A (1975) 1655.
25. S. Foruno, K. Izuik. K. Ono, T. Kino, J. of Nuclear Materials, 133-134 (1985)400.
26. P.Rozenak, J. of Alloys and Compounds, (2005) in Press.
27. G.Lu, Q. Zhang, N. Kioussis, E. Kaxiras, Physical Review Letters, 87 (9) (2001) 095501-4.
28. G.M. Bond, I.M.Robertson, H.K. Birnbaum, Acta Metall., 35 (9) (1987) 2289.
29. J. Mao, R.B. Mclellan, J. of Physics and Chemistry of Solid, 62 (2001) 1285.
30 C.E Buckley, H.K. Birnbaum, J. of Alloys and Compounds, 330-332 (2002) 649.
31. J.W. Watson, Y.Z. Shen, H.Meshii, Metal. Trans.A, 19A (1988) 2299.
32. D. Milcius, L.L.Pranevicius, C.Templier, J. of Alloys and Compounds, 398(2005) 203
33. W. Eichenauer, K. Hattenbach, A. Pebler, Z. Metallk., 52 (10) (1961) 682.
34. S. Matsuo, T. Hirata, J. Japan Inst. Metals, 43 (1979) 876.
35. M. Ichimura, M. Imabayashi, M. Hayakawa, Japan Inst. Metals, 43 (1979) 876.
36. K. Papp, E. Kovacs-Csetenyi, Scripta Metall., 15 (1981) 161.
37. R.A. Outlaw, D.T. Peterson, F.A. Shmidt, Scripta Metall., 16 (1982) 287.
38. H. Saitoh, Y. Iijima, H. Takana, Acta Metall., 42 (7) 2493.
39. R. Braun et al., in Hydrogen Transport and Cracking in Metals, ed. A.Turnbull. Inst. of Mater., Teddington (1994) 280.

Aluminum Alloys
for Transportation, Packaging, Aerospace and Other Applications

Alloy Characterization

Aluminum Alloys for Transportation, Packaging, Aerospace, and Other Applications
Edited by Subodh K. Das, Weimin Yin
TMS (The Minerals, Metals & Materials Society), 2007

Microstructural Characteristics during Hot Forging of Al-Mg-Si Alloy

Yong-Nam Kwon, Y. S. Lee, J. H. Lee

Materials Processing Research Center, Korea Institute of Machinery and Materials
66 Sangnam-dong, Changwon, 641-010, Korea

Keywords: hot forging, Al-Mg-Si alloy, power dissipation map

Abstract

Thermomechanical behavior of Al-Mg-S i alloys was studied to in vestigate th e ef fect of microstructural featu res such as pre-exis ting s ubstructure and distribu tion of particles for a successful hot forging. The controlled com pression tests were carried out to find out how the alloy responds to temperature, strain amount and strain rate. Then hot forging of Al-Mg-Si alloys was carried out and an alyzed by the com parison with th e com pression tes ts. M icrostructural features after forging were discussed in term s of the thermom echanical respon se of Al-Mg-Si alloys. The deformation of Al-Mg-Si at the elev ated temperature brought the recovered structure on m ost conditions as well reported before. Ho wever, abnorm ally large grains could be developed due to a specific forging conditions such as friction condition between die and forging stock, which leaded to a huge gr ain growth. These undesirable m icrostructural variations could give a rise a degradation of mechanical properties, especially fatigue strength.

Introduction

Weight reduction for autom otive has m any advantag es su ch as better fuel efficiency, driving performance and safety as shown in Fig. 1. One of practical and easy w ays to m eet the dem and for weight reduction is the use of more Al parts. Therefore, there have been many Al components applied in passenger cars for the las t decade. Lo wer control arm is one of the m ost vigorously substituted parts from steel in to Al alloys. Typically two types of Al alloys are being used for lower control arm fabrication; wrought alloys of 6xxx series and cast alloys. Usually, wrought Al parts fabricated by hot forging have better m echanical properties over cast Al alloys. The alloys such as 6061 and 6083 Al alloys are typical alloys for Al control arms.

Various aspects of hot working have been studi ed to elucidate the optim ized processing window and to improve the quality of the final product. And most researches on hot working of Al alloys have been f ocused on the hot extrusion process since it m ight have bigger m arket than hot forging. However, it m ight be interesting to understand how the proce ssing variables such as strain and strain rate coupled with f orging te mperature influence hot fo rging process since the forged parts usually have m ore complex shapes than the s imple ex truded bar. F ig. 2 represents specifically how forged structur e can be developed when the forging conditions are not m et properly. In fact, abnormally gr own grains would be obtained under a certain conditions, which microstructure might have a detrimental effect on fatigue and fracture characteristics.

In the present study, therm omechanical behavior of a commercial Al-Mg-Si alloy was studied to investigate the effect of processing conditions for a successful hot forging. The controlled compression tests were carried out to find out how the alloy responds to tem perature, strain amount and strain rate. Then hot forging of Al-Mg-Si alloys was carried out and analyzed by the comparison with the co mpression tests. Micros tructural features after forg ing were discussed in terms of the thermomechanical response of Al-Mg-Si alloys.

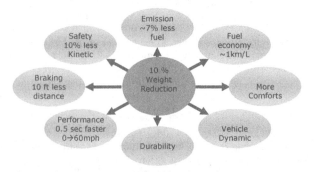

Figure 1. Advantage of weight reduction for usual passenger car[ ]

Figure 2. A f orged part cross se ction illu strating how the m icrostructural va riation could occu r when the processing conditions are not m et we ll. At a certa in area like right bottom corner could be observed the finely recrystallized structure.

Experimental Procedures

A commercial grade 6061 alum inum alloy with m odified with a slight amount of M n was used. The alloy has been continuously cast and homogenized at 550 °C for 6hrs. Also, the extruded billet was u sed for the com parison purpose. Als o, pre heat treatm ent be fore com pression tes ts was carried out to change the microstructural features such as size of precipitate.

A series of com pression tes ts were carried ou t in the tem perature range of from 350 to 520 °C with the strain rate of 0.001~50/s by using Gleeble m achine. Geom etry of test sample was a 12mm height cylindrical bar with the d iameter of 10mm . All sa mples were hea ted up to tes t temperature at the speed of 100 °C /min and hol d for 5 m ins to prevent a volum etric change due to thermal expansion before com pression. Tests we re performed to over the strain of 1.0. Stress-strain ra te c urves at the various te mperatures were plotted using the co mpressive flow curves. The strain r ate sensitivity was deter mined subsequently from stress-strain ra te cu rves for further analysis. A forging test was carried out by using crank press with the deformation speed of about 10/s. Forging variables like temperature and pre-heat treatments were controlled.

Results and Discussion

Typical flow curves of Al-Mg-Si alloy with the variation of temperature are shown in Fig. 3. The variation of strain rate did not give many differences in the pattern of flow curves. The steady state flow curves represent that recovery dominated deformation behavior is dominating under the current test conditions. Steady state flow could be expected due to the high stacking fault energy of Al and its alloys. The softening behavior was also observed when the strain rate gets lowered irrespective of test temperature. Even in the softening flow curves, the steady state would be found after a small strain level.

Stress-strain rate relation with the variation of temperature is plotted in Fig. 4 for the cast billet. The stress level of each curve was determined at the strain of 0.6, which is believed that the deformation was relatively stabilized. The overall strain rate sensitivities ranged from 0.1 at 400°C to 0.14 at 520 °C, which illustrates that deformation behavior is a recovery dominant process indirectly. Stress-strain rate curves for extruded billet did not show much difference with those of cast one. The activation energy at the test temperature range was observed to be around 130kJ/mol, which corresponds to Al self diffusion energy.

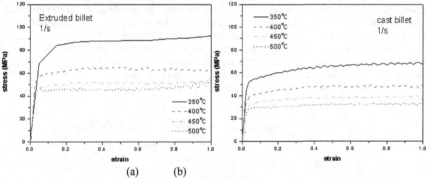

(a)   (b)

Figure 3. Flow curves of Al-Mg-Si alloy with temperature variation (a) extruded and (b) cast billets.

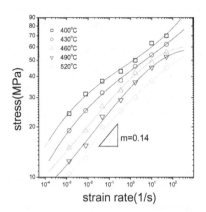

Figure 4. Stress-strain rate curves with the variation of temperature.

Fig. 5 shows the optical m icrographs of 70 % compressed sam ple at the tem perature of 520 °C with variation of strain rate. Ev en though the flow curves did not give any clue of nucleation of recrystallized grain, it was possible to observe a small grain occurring along the grain boundaries of compressed billet. In the co mpression test, there were no a bnormally grown grains observed under test conditions. However, the undesirab le m icrostructure of abnorm al coarse grains could occur when the deformation condition was not op timized as shown in Fig. 2. Therefore, the coarse grains in Fig. 2 might be originated from the recrystallized grains which could be growing rapidly during solution treatm ent that is given to forging product necessary mechanical strength. It is po ssible to assume that th e te mperature difference between forgin g and solution treatm ent would control how fast the newly recrystallized grain could be growing in solutionizing temperature.

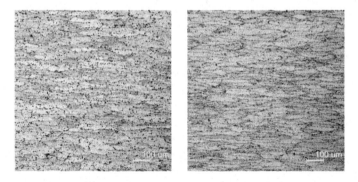

Figure 5. Optical m icrographs of 70 % compressed sample at 520°C with strain rate of (a) 10^{-3}/s and (b) 10/s

Various m icrostructural features could appear in a forged product since ev ery section of forged part dif fers in str ain, str ain ra te and tem perature rise. It is hard to pre dict th e m icrostructural evolution with a sim ple approach. In order to understand the effect of forging param eters, die forging was carried out with a certain geom etry which could not be explained in detail. Since a strain rate would be fixed due to press speed, the temperature of hot forging usually tends to be a processing variable for hot forging process.

Fig 6. show s the forged m icrostructures of cast billets with variation of tem peratures. Forging temperature here m eans the billet temperature be fore forging. And die tem perature was fixed at the tem perature of 220 ±30°C. Each row repres ents three dif ferent cross sections with different strain amount. And column corresponds to the sa me processing temperature. Forging of the first column was done at 500°C by following solutionizing at 550°C for 3hrs. When solution time and temperature were decreased, the microstructure gave improved structure with less level of coarse grains. W hen both temperature and tim e for solu tion treatm ent are decrea sed, most areas of forged cross section left without coarse grained structure. In this case, recovered structure prevailed as observed in the com pressed sam ple. Th erefore, it is possible to conclude that hot forging of a commercial grade 6061 Al alloy was found to be the recovery dom inated process. However, th e accum ulated energy in forged billet seem s to be releas ed by nucleation of fine recrystallized grains along the grain boundary or by driving ra pid grain growth during the following solutionizing step. These phenom ena are closely related with internal state how Mg and Si atoms reside in m atrix even though it did not get into in detail. This precipitate kinetics related behavior should also be addressed in deta il in order to elucidate the optim um processing window for Al hot forging.

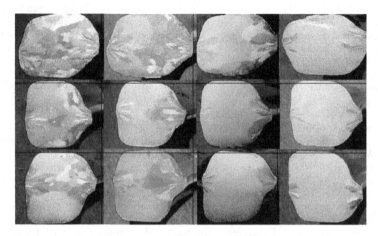

Figure 6. F orged m icrostructures w ith variation of tem perature. Each row represen ts different cross sections. And each column corresponds to the same processing temperature

Conclusion

In the present study, a series of compression tests for Al-Mg-Si alloys perform ed to find out the thermomechanical response under the different deformation conditions. Also, forging experiment has been done using m odel geometry with a large variation of strain. From the present study, it was shown that a ho t forging of Al alloy shou ld be contro lled carefully to g et the succes sful results. Temperature for forging and following heat treatment should be selected carefully since it m ight determ ine the f inal m icrostructure o f forged part, which is closely related with mechanical properties.

Acknowledgement

This work was supported by grant No.104-01- 03 from the Regional Technology Innovation Program of the Ministry of Commerce, Industry and Energy(MOCIE) of Korea..

References

1. F.J. Hum phreys and M. Matherly, Recrystal lization and Related An nealing Phe nomena, Pergamon, (1995).
2. R. D. Doherty, etc , Mater. Sci. & Eng. A238 (1997), 219.
3. H. J. McQueen and O. C. Celleiers, (1997) Application of hot wo rkability studies to extrusion processing, Canadian Metall. Quart., Vol. 36, No. 2., pp.73-86.
4. Y. V. R. K. Prasad and S. Sasidhara, (1997) Hot working guide, ASM international, Materials Park, OH.

Aluminum Alloys
for Transportation, Packaging, Aerospace and Other Applications

Alloys
Mechanical
Behavior

Aluminum Alloys for Transportation, Packaging, Aeropsace, and Other Applications
Edited by Subodh K. Das, Weimin Yin
TMS (The Minerals, Metals & Materials Society), 2007

CHARACTERIZATION OF THE EVOLUTION OF THE PROPERTIES OF ALUMINIUM ALLOYS

Christophe Thiebaut, Serge Contreras, Laurence Durut, Jean-François Mariage, Thierry Vauzelle

CEA; Centre de Valduc; Is sur Tille, 21120, France

Keywords: aluminium, 7020, container, aging, breakdown pressure

Abstract

Aluminium alloys are now widely used in quite every major applications: automobile, aircraft, consumer goods. They have a lot of advantages: they are light, have good mechanical properties, are easily forged and machined and some are compatible with hydrogen use. So we have begun studies of the use of aluminium alloys for hydrogen storage under pressure, more precisely with Al-Zn-Mg alloys. These are easily weldable.
Most of the results that are in the literature consist in MIG welds. After forming and welding by electron beam technique, we make a heat treatment. Experiments are then made in order to compare the resistance to breakdown of the storage container by hydraulic test. This can be done at different aging of the containers. Some hypotheses are made to try to explain the results. Further characterization will be necessary to understand the phenomena which are seen.

Introduction

The mechanical properties and the corrosion resistance of the 7000 series alloys have been studied a lot. This is explained by their low density, their rather good weldability, their resistance to corrosion and their ability to increase of the strength by age-hardening. There are mainly used for constructional purposes [1] but also for aerospace application [2, 3, 4]. These alloys age-harden because of the formation of Guinier-Preston zone which leads then to $MgZn_2$ precipitates.
Al-Zn-Mg alloys, which are of the 7000 series, have good mechanical properties: the yield stress varies from 230 MPa to 350 MPa and the maximum stress is between 340 MPa and 450 MPa without copper addition whereas the yield stress lies between 350 and 720 MPa and the maximum stress between 430 and 780 MPa when copper is added to the composition. This kind of alloys can be obtained in the T652 state through steps of forging and heat treatments. The parts of the piece are then machined to the final dimensions, welded together by electron beam technique with three zones: one with one pass of the beam (1PZ), one with two passes (2PZ or overlap) and another one with the end of the weld (SZ or slope down). They go then through a solutionizing heat treatment followed by a two steps aging heat treatment which leads to a T6 state. Most of the results published concern MIG or TIG welding [5, 6]. Only a few studies integrate electron beam welding [7]. In our case, we will do containers which will be studied for hydrogen storage. So its interaction with the alloy can induce some differences.

The parts are then aged with different conditions, different heat treatments. There remaining resistance are then characterized globally by a break test which yields to the failure of the container when we increase progressively the inside pressure.

Aluminium Alloys

Aluminium alloys can be very different. There are age hardening alloys and non age hardening alloys. In table 1, we have listed the different metallurgical states of the 7000 alloys that can be purchased commercially. In addition to the data given in table 1, the following indications can be useful: the H states can be subdivided: HXY (Y = 2, 4, 6, 8, 9: mild → extra hard); the T states can be subdivided also: TXY (Y = 1 to 6: soft tempering → hard tempering). The terminology TX51 means that the stress is relieved by tensile deformation, TX52 stands for stress relieving by compression and TX53 is for stress relieving by heat treatment.

Table 1: Different Metallurgical States of the 7000 Alloys

Base states	commentaries	subdivisions	definition
F	rough of elaboration	/	
O	annealed	/	
H	cold worked	H1	cold working
		H2	cold working + partial annealing
		H3	cold working + annealing of stabilisation
W	quenched - non stabilised		
T	heat treated	T1	cooling after hot transformation + natural ageing
		T3	solution treated + quenching + cold working + natural ageing
		T4	solution treated + quenching + natural ageing
		T5	cooling after hot transformation + tempering
		T6	solution treated + quenching + tempering
		T7	solution treated + quenching + over-ageing
		T8	solution treated + quenching + cold working + tempering
		T9	solution treated + quenching + tempering + cold working
		T10	T5 + cold working
		T11	cooling after hot transformation + cold working + natural ageing
		T12	cooling after hot transformation + cold working + tempering

The solution treatment is made so as to obtain the partial dissolution of the metallic phases in the alloy. It follows that after this treatment we have an homogenization of the solid solution. The quenching permits to achieve a solid solution in over saturation at room temperature (out of equilibrium state): Zn and Mg in particular are not in a stable state. So through natural or artificial ageing, there is a decomposition of the metastable solution to equilibrium by hardening precipitation: this is done through the formation of Guinier-Preston zones, which are solute atomic clusters, coherent with the matrix. They induce a lattice distortion which is mainly elastic deformation. These zones are very fine (< 5 nm). When ageing is artificial, there is an acceleration of the natural age hardening, which induces an increase of the final hardening level: the Guinier-Preston zones grow and then if temperature is high enough, they dissolve. This is followed by precipitation of MgZn which are metastable, semi-coherent with the matrix. At the end only, MgZn2 precipitates are formed which are stable but incoherent with the matrix [8].

If the age hardening treatment is followed, we can obtain a T7 state: there is a coalescence of the precipitates and a softening of the mechanical properties of the alloy.

As it has been described in the introduction, copper has a strong effect on the mechanical properties for instance. But it has also a strong effect on corrosion resistance. A good compromise can be made by using 7020 alloys which are age hardening, but retains a good corrosion resistance. Furthermore, there are easily weldable.

Table 2 gives the composition of the normalized 7020 alloy, the composition which has been asked by CEA and the typical value of the products which has been received. 7020 alloy is characterized by the presence of Zn and Mg which can then form $MgZn_2$ precipitates with aging whereas it is done by time or temperature.

Table 2: Composition of 7020 Alloys Given by the EN AW, the CEA Modifications and the Values for a Typical Batch

	Zn	Mg	Cr	Mn	Fe	Si	Cu	Zr	Ti	Z< 90
EN AW 7020 composition	4.0-5.0	1.0-1.4	0.1-0.35	0.05 -0.5	< 0.4	< 0.35	< 0.20	0.08-0.35		
CEA 7020 composition	4.4-4.9	1.1-1.4	0.15-0.35	0.05 -0.2	<0.3	<0.3	< 0.15	0.08-0.20	<0.0 6	< 0.15
Typical batch	4.52	1.18	0.26	0.11	0.12	0.13	0.08	0.12	0.04	0.05

Experimental

We will describe here how the samples are prepared from the foundry down to the final treatment of the welded parts. The Al-Zn-Mg alloys that we have purchased followed the classical metal forming way for this kind of alloys which can be obtained in the T652 state. There were obtained by: casting first an ingot of several thousands kilograms. The ingot was then cut into several billets, which were themselves cut into several pieces of around 50 kilograms. These pieces were then forged under a press through a progressive rotation. The forging induces a microstructure which is elongated and fibre like. The pieces are then cut into the final desired length. The process is shown on figure 1.

The alloy is then heat treated at 465°C for 4 hours (solution treatment) and water quenched. The stress is relieved by a compression in the range of 1 to 5%. Then a double tempering heat treatment is made at 100°C for 8 hours, followed by 130°C for 24 hours. Thus we obtain a 7020-T652 alloy. This has been extensively studied by numerous teams of searchers [9, 10]. This fabrication is fully characterized through chemical analysis (results given in table 2), optical metallography, ultrasounds characterization (detection of defects smaller than 1.6 microns), electrical conductivity, mechanical properties (tensile test and hardness).

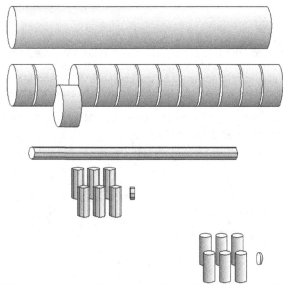

Figure 1: process of fabrication of a typical aluminium alloy

The samples are then machined to the desired dimensions. In our case, we have made two hemi spherical parts. These parts have been welded together with an electron beam machine by rotation of the parts. The current is progressively increased and the parts make a 360° rotation. After this first round, the welding goes on for around 90° and then the current decreases. The sketch of the welding is presented on figure 2.

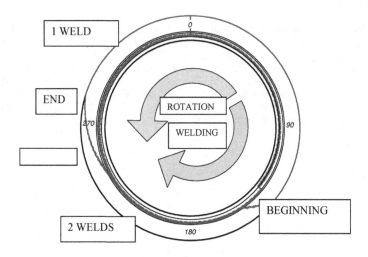

Figure 2: sketch of the welding of the parts

It is well known that the different parameters of the welding are important: accelerating voltage, beam current, welding speed and chamber pressure. All these parameters can induce changes in the microstructure and the chemical composition of the weld itself. For Al-Zn-Mg in particular, Zn and Mg can evaporate. The segregation can also be quite different compared to the base metal. The precipitation state is thus very different compared to the base metal. So in the process used for this study, we have decided to do a second double tempering heat treatment but we will not have a T652 state, because it is not possible to perform a deformation of 1 to 5% between a solution treatment and the final heat treatment.

We have thus performed the following steps after welding:
- solution treatment at 465 ± 10 °C for half an hour, in a closed oven, under partial nitrogen atmosphere,
- quenching with pulsed nitrogen (Quench speed > 0.5 °C/s), in the same oven, down below 300°C, and then decrease of the temperature to 50°C, when the door of the oven is opened so as to achieve ambient temperature,
- natural ageing: 5 days minimum at 20°C, in air atmosphere,
- artificial ageing through a double tempering heat treatment in a closed oven, under partial nitrogen atmosphere: first at 100 ± 5 °C for around eight hours followed by twenty four hours at 130 ± 5 °C,
- then at the end, a low cooling under furnace, at a speed comprised between 2 to 3.5 °C/min. The sketch of the heat treatment is given on figure 3.

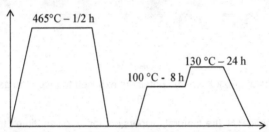

Figure 3: sketch of the double tempering heat treatment of the welded parts

If we compare the two cycles of the aluminium alloys as purchased and the parts after welding, it can be seen that the main difference is that there is no deformation after the solution heat treatment. The other differences concern the solution treatment itself which is longer for the ingots as purchased (four hours versus half an hour for the welded parts) and the quench which is faster for the purchased alloys. Last the maximum temperature of the tempering temperature is 5° below (130 versus 135).

Results and Discussion

The samples are used for a research concerning development of containers for hydrogen storage. As a matter of fact, the problem faced by the designers of hydrogen powered devices is precisely hydrogen storage. The idea in this part of the research is precisely to see if this kind of aluminium alloys which are light, have good mechanical properties and have also good corrosion resistance would be suitable for this storage application.

The research began a few years ago: samples have been filled with hydrogen gas under different pressures. After varying times, there are emptied and tested until breakdown: the pressure is progressively increased until a leak or a break of the container which induces a decrease of the pressure.
We have also made this kind of test for new containers. All the data are compiled in figure 4.

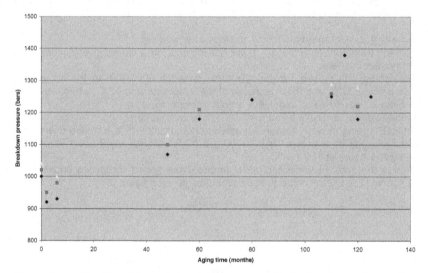

Figure 4: results of breakdown for containers exposed to hydrogen gas for varying times

First thing to notice is the natural dispersion for a given time: the breakdown pressure varies from 50 bars to 200 bars, which is between 5 to 15%. This is quite a rather large discrepancy but it can be explained by the inherent heterogeneity of the purchased material and then the small differences in the process which come at every step, especially during the differing heat treatments.

Second thing to notice is the evolution of the breakdown pressure versus time: it increases continuously from 0 to 130 months. This can be explained different ways: 1) the mechanical properties are not at the maximum after heat treatment. But the heat treatment which has been chosen is the nominal T6 heat treatment which gives normally the maximum of the resistance. The only difference as has been pointed out in the preparation of the sample is that there is no plastic deformation between the solution treatment and the tempering treatment, and the small differences in the thermal cycles. But can it induce such an evolution? We have to keep in mind that the latest heat treatment temperature is 130°C which is largely above room temperature which is the temperature where the containers are stored. 2) There can be a chemical interaction of hydrogen with the alloy: hydrogen would induce an increase of the mechanical properties versus time and the more the container contains hydrogen, the better the mechanical properties. 3) There can be a mechanical effect of the pressure of hydrogen which could induce dislocation on the micro scale, which then would induce an increase of the mechanical properties.

So, we have to plan some other experiments to try to understand what we have seen during this study: for the first point, we will do some container which will be stored without pressure in order to determine the effect of the pressure on the breakdown pressure. We also have to make some variations on the heat treatment parameters in order to see if it changes the breakdown pressure. It would be also very interesting to make some measurements for larger times (for example 200 months) but we need to have more samples. We will also perform hardness measurement on a part of the container which has not seen hydrogen and study its evolution versus time. For the second and the third points, we have to characterize at the smallest possible scale, i.e. with Transmission Electronic Microscope (T.E.M.) and chemical micro probe in order to see if there is a diffusion of hydrogen or if there is an increase of dislocation. Furthermore, we will do optical metallography. In the end, we will also characterize the weld versus the base metal in terms of grain size, precipitates and chemical concentration.

Conclusion

We have reported first results of the evolution of the breakdown pressure of hydrogen containers for various times of storage. We have seen that there is a natural dispersion of the values ranging between 5 and 15%. The pressure breakdown increases continuously versus aging time from around 1000 bars for new containers up to nearly 1400 bars after 130 months. We have seen that this can be explained by either the initial heat treatment parameters are not optimal for a T6 alloy or there is an interaction of the pressure or the gas with the container which could explain this evolution. A set of experiments has been descried which could give some elements for explaining the evolution.

References

1. E.H. Hollingsworth, H.Y. Hunsicker, *Corrosion* (Metals Handbook, 9th edition, vol. 13, ASM, Metals Park, OH 1987) p.583
2. K.V. Jata, E.A. Starke Jr, "Fatigue crack growth and fracture toughness behaviour of an Al-Li-Cu alloy", *Metallurgical Transactions, A*, 17 (1986), 1011-1026
3. G. Bussu, B.D.Dunn, "ESA approach on the prevention of stress-corrosion cracking in spacecraft hardware", (Paper presented at the Joint ESA-NASA Space Flight Safety Conference, June 2002, ESA/ESTEC, Netherlands)
4. A. Heinz, A. Haszler, C. Keidel, S. Moldenhauer, R. Benedictus, W.S. Miller, "Recent development in aluminium alloys for aerospace applications", *Materials Science and Engineering*, A280 (2000), 102-107
5. T.K.Chan, R.F.D. Porter Goff, "Welded aluminium alloy connections: test results and BS8118", *Thin walled structure*, 36 (2000) 265-287
6. H. Cordier, M. Schippers, I. Polmear, "Microstructure and intercrystalline fracture in a weldable Al-Zn-Mg alloy", *Z. Metallkde*, 68 (1977) 280-284
7. C.H. Lee, S.W. Kim, E.P. Yoon, "Electron beam welding characteristics of high strength aluminium alloys for express train applications", *Science and Technology of Welding and Joining*, 5 (2000) 277-283
8. M. Nicolas, J.C. Wernskiold, A. Deschamps, F. Bley, F. Livet, J.P. Simon, "Study of precipitate microstructures in the heat-affected zones of a welded Al-Zn-Mg-alloy", (Paper presented at Euromat 2001, Riminy, Italy).

9. A. Deschamps, F. Livet, Y. Brechet, "Influence of predeformation on aging in an Al-Zn-Mg alloy-I. Microstructure evolution and mechanical properties", *Acta Mater.*, 47 (1999) 281-292

10. X.J. Jiang, B. Noble, B. Holme, G. Waterloo, J. Tafto, "Differential scanning calorimetry and electron diffraction investigation on low-temperature aging in Al-Zn-Mg alloys", *Metallurgical and Materials Transactions A*, 31A (2000) 339-348

Aluminum Alloys for Transportation, Packaging, Aeropsace, and Other Applications
Edited by Subodh K. Das, Weimin Yin
TMS (The Minerals, Metals & Materials Society), 2007

EFFECT OF AGING TREATMENT ON THE MECHANICAL
PROPERTIES OF THIXOEXTRUDED 7003 AL WROUGHT ALLOY

Young-Ok Yoon, Hoon Cho, Shae K. Kim and Hyung-Ho Jo

KITECH(Korea Institute of Industrial Technology); 7-47 Songdo-dong, Yeonsu-gu, Incheon,
406-840, Korea

Keywords: Thixoextrusion, Isotropy, 7003 Al Wrought Alloy

Abstract

In the present study, the influences of thixoext rusion param eters, such as isotherm al holding temperature of billet, initial r am speed and bearing len gth, on m echanical pr operties of thixoextruded 7003 Al wrought alloy were investig ated. A lso, the effect of aging treatm ent on the mechanical properties of thixoextruded 700 3 Al wrought alloy was investigated. The study for thixoextrusion of 7003 Al wrought alloy was carri ed out with respect to isotherm al holding temperature and tim e during the par tial remelting, especially in the low liquid f raction ($f_L<0.3$). The liquid fraction and average grain size with respect to isotherm al holding tim e were alm ost uniform. The m aximum extrusion pr essure of the thixoextrusion was thr ee times lower than that of the hot extrusion. The tensile and yield strengthes of the thixoextruded bar before T5 agin g treatment were lower than those of the hot extruded bar, while the elongation value of the thixoextruded bar was higher than that of the hot extruded bar. However, the tens ile and yield strengthes of the thixoextruded bar af ter ag ing trea tment were sim ilar to hot extruded bar. Therefore, their low tensile a nd yield strengthes could be imp roved through appropriate heat treatments.

Introduction

The 7003 Al wrought alloy has been used for stru ctural applications wh ere high m echanical strength is needed and in the au tomotive indu stry. Howev er, it gen erally a llows low extrus ion speed and low extrudability index when extrude d conventionally and also causes rather high extrusion pressure. Thixoextrusion, one of the thixoforming processes, has advantages in point of improving high productivity, reducing extrusion pr essure and obtaining ho mogeneous internal microstructure compared with conventional extrusion processes. [1-5]

For the thixoextrusion process, the reheating cond itions to obtain thixotropic m icrostructure are very im portant. Moreover, the re heating of th e bille t in the se misolid state as quickly and homogeneously as possible is the m ost im portant part of the process. Conventional electric furnace heating invo lves long heating tim e a nd cannot be contro lled to obtain unifor m temperature distribution so that the induction heating m ethod is usually used in the reheating process. Kapranos et al. com pared num erical re sults with the tem perature of five points in experiments with heating, soaking, and power-off stages. [6] In a study on the reheating process, Akbas and Turkeli investigated the changing pr ocess of the globulariza tion according to the heating temperature of 7075 Al wrought alloy. [7]

However, previous papers were extensively fo r thixo casting in high li quid fraction. No paper about m icrostructural evoluti on in the low liquid fraction (f $_L<0.3$) for thixoextrusion has been reported. For thixoextrusion, the low liquid frac tion should be achieved and also the liquid fraction and average grain size should be uniform according to the isothermal holding time at the desired low liquid fraction.

This paper m ainly discusses m icrostructural e volution with respect to isotherm al holding temperature and during the partial rem elting for 7003 Al wrought alloy withou t addition al pretreatment. Also, The present study discussed extrudability improvement for 7003 Al wrought alloy by thixoextrusion, with em phasis on con trolling thixoextrusion parameters, such as initia l ram speed, die bearing length and extrusion temperat ure of billet in semiso lid state. The results of thixoextrusion experiments about microstructures and extrusion pressures were compared with conventional hot extrusion results.

Experimental Procedures

The material used in this study was commerc ially available 7003Al wrought alloy. The 7003 Al wrought alloy (54 mm diam eter & 210 mm length) has been m anufactured by m elting in a high-frequency induction furnace. Homogenization treatment was carried out at temperatures of 480~ for 8hrs. The liquid fraction (f $_L$) at any tem perature (T) within liquid-solid two phase range is usually given by the Sc heil equation assum ing that the liquid is com pletely homogenous and no diffusion occurs through the solid, where T $_M$ is the m elting temperature of the pure metal, TL is the liquidus tem perature of the alloy and ko is th e equ ilibrium distribution coefficien t. In th is Scheil equation, the value of T$_M$ is about 660~, the value of T $_L$ is about 655~ and 0.193 ko were taken, based on the DSC result.

The billet was m achined to th e size of 48 mm diam eter and 20 mm length and m icrostructural evolution of specim ens in the partially rem elted semisolid state was carried out in an induction reheating equipment at the desired liquid fraction for tim es varying from 0 to 30min, after which the specimens were quenched into water imm ediately. The isotherm al holding time was decided based on actual extrusion tim e, therefore m aximum isotherm al holding tim e wa s 30m in for observation of m icrostructural evolution by long holding tim e. This is because the uniform average grain size and liqu id fraction accord ing to th e isotherm al h olding tim e were ve ry important f or thixoex trusion in term s of actual ex trusion tim e. The char acteristics of microstructural evolution, such as liquid fraction, average grain si ze and distribution of globular grain, were evaluated by image analysis system (Image-Pro Plus) and optical microscope.

The billet was partially rem elted inside the s leeve by cartridge-type he aters. The te mperature of the billet in the sleeve w as monitored through a K-type thermocouple located at the center of the top surface of the billet. After the required tem perature was r eached, the ram was horizon tally moved for thixoextrusion process. The load and the stroke in extrusion were m easured using a load cell an d recorded with a pe rsonal computer. Cylindrical bars with the ex trusion ratio of 11 were extruded by using the 20ton horizontal thixoextrusion apparatus. The ram speeds were 5 mm /s and 15 mm/s. The angle of the thixoextrusion die was 2 α=90° and the die temperature was 500°C. The die bearing lengths were 3 mm and 7 mm. The two-step aging treatm ent (T5) for extruded bar and thixoextruded bar was performed at 90 °C for 6hr and subsequently 180 °C for 7hr without solution treatm ent. The experim ental conditions of extrusion process in this study were shown in Table 1.

Table 1. Experimental conditions for the extrusion process.

	Hot extrusion	Thixoextrusion			
	Case1	Case2	Case3	Case4	Case5

Extrusion temp. (°C)	450	624	640	624	624
Die bearing length (mm)	7	7	7	3	7
Initial ram speed (mm/sec)	3	5	5	5	15
Die temp. (°C)	250	500	500	500	500
Extrusion ratio	11	11	11	11	11

Result and Discussion

As-quenched m icrostructures after isothermal holding for 0m in at (b) 624~, (c) 640~ and (d) 645~ were shown in Fig. 1 with the hom ogenized specimen. With increasing temperature, partial remelting was found to start at eutectic phase. The isotherm al holding tim e was 0m in at all temperatures. As shown in Fig. 1, the total liquid fraction and average grain size were slightly increased with increasing isothermal holding temperature.

Fig. 2 shows the m icrostructures of the specim ens holding at 640 °C for 0, 2, 5, 10 a nd 30 m in. The total liquid fraction was uniform regardless of increasing isotherm al holding tim e. But, entrapped liquid inside the solid particles was slightly decreased. When holding time was 0min, the prim ary grains becom e globular obviously without rosette m orphology. After holding for 5 min, the prim ary grains begin to turn into globula r ones. The grains grow and turn into globular ones through the m ovement of the grain boundarie s. The uniform average grain size and liquid fraction according to isotherm al holding time were very important for thixoextrusion in term s of actual extrusion time.

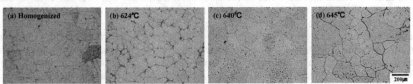

Figure 1. Microstructures of 7003 Al wrought alloy with respect to isothermal holding temperature after partial remelting for 0min.

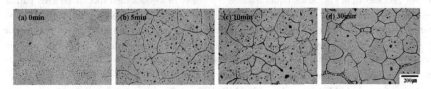

Figure 2. Microstructures of 7003 Al wrought alloy with respect to isothermal holding time after partial remelting at 640~.

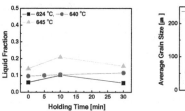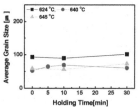

(a) total liquid fraction   (b) average grain size

Figure 3. Results of microstructure evaluation of 7003 Al wrought alloy with respect to isothermal holding time after partial remelting at the reheating rate 60˜/mm.

The results from the m icrostructural evolutions as a function of isotherm al holding tim e for thixoextrusion were summarized in Fig. 3. Fig. 3(a) sho ws the liqu id f raction with respe ct to isothermal holding tim e at the given tem peratures. The liquid fracti on was uniform with increasing isothermal holding tim e. Fig. 3 (b) shows the av erage grain size was almost uniform with increasing isothermal holding tim e. The fac t that the liquid fraction and average grain siz e with respect to isothermal holding tem perature and time were almost uniform is very im portant and considered useful for thixoextrusion in term s of process control such as billet tem perature control and actual extrusion time.

Fig. 4 shows the comparison of m aximum ex trusion pressu res in ho t e xtrusion an d thixoextrusion. The m aximum extrusion pressure of thixoextrusion was three tim es lower than that of conventional hot extrus ion. The m aximum extrusion pressu re of the thixoextrusion was 130.6MPa (Case 2) and the m aximum extrusion pressure of the hot e xtrusion was 416.5MPa (Case 1). Because of the low flow stress of th e bille t in the sem isolid sta te, the m aximum extrusion pressure necessary fo r the thixoextrusion wa s lower than that of conventional hot extrusion. Owing to the lubrication effect of th e liquid phase and the low extrusion pressure, products of com plicated cross sect ional profiles and thin wall-thi ckness could be thixoextruded rather easily. As shown in Fig. 4, the m aximum extrusion pressure was decreased from 130.6MPa (Case 2) to 88.7MPa (Case 3) with in creasing thixoextrusion tem perature from 624˜ to 640˜. As shown in Fig. 3, the liquid fraction value of as-quenched specim en a fter partial remelting for 0min at 640˜ was about two tim es higher than that of as-quenched specim en after partial remelting for 0min at 624°C. It was pointed out that the extrusion temperature dependence of the m aximum extrusion pressure was large a nd the influence of extrusion tem perature on the improvement of extrudability was remarkable in thixoextrusion process. The maximum extrusion pressure of the thixoextruded bar after isotherm al holding for 0m in at 624 °C was slightly increased from 124.8MPa (Case 4) to 130.6MPa (Cas e 2) with increasing die bearing length from 3 mm to 7 mm. The m aximum extrusion pressure of th e thixoextruded bar after isotherm al holding for 0min at 624°C was slightly decreased from 130.6MPa (Case 2) to 130.4MPa (Case 5) with in creasing in itial ram speed f rom 5 mm/s to 15 mm/s. But, the m aximum extrusion pressures with respect to initial ram speed were practically equal.

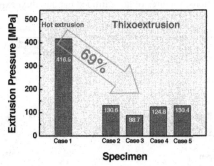

Figure 4. The maximum extrusion pressures of 7003 Al wrought alloy at the given conditions.

The m icrostructures of longitud inal section a t the each po sition of th e hot extru ded bar and thixoextruded bar of 7003 Al wrought alloy w ere shown in Fig. 5. A s shown in Fig. 5(a), the elongated grains which have the axisymmetry with extrusion direction w ere generally observed during hot extrusion process. Fig. 5(b) shows the microstructures of the thixoextruded bar after isothermal holding for 0min at 624°C with the initial ram speed of 5 mm/s. The microstructures of the thixoextruded bar were isotropic.

Fig. 6 shows the variation in tensile properties of the before and after aging treatm ent of the hot extruded bar and thixoextruded bar. As shown in Fig. 6(a), the te nsile and yield strength values of the hot extruded bar before aging treatm ent were 335MPa and 225MPa, respectively (Case 1). Before agin g tre atment, the tens ile and yield st rength values of the thixoextruded bar were 238MPa and 118MPa, respectively (Case 2). The tensile and yield strengthes of the thixoextruded bar were lower than those of the hot ex truded bar, while the elongation value of the thixoextruded bar w as higher than that of the hot extruded bar. However, as shown in Fig. 6(b), the tensile and yield st rengthes of the thixoextruded bar after aging trea tment were s imilar to hot extru ded bar. Therefore, their low tens ile and yield strengthes co uld be im proved through appropriate heat treatments.

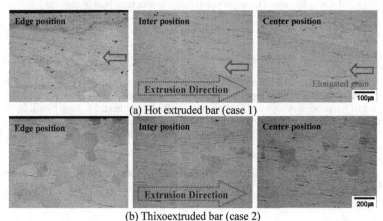

(a) Hot extruded bar (case 1)

(b) Thixoextruded bar (case 2)

Figure 5. Microstructures of hot extruded bar and thixoextruded bar.

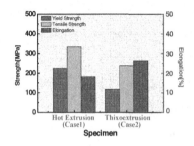 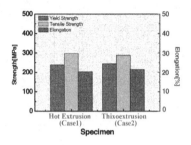

(a) Before aging treatment (As extruded)   (b) After aging treatment
Figure 6. Mechanical properties Microstructures of hot extruded bar and thixoextruded bar.

Conclusion

In the present study, the influences of thixoext   rusion param eters, such as isotherm al holding temperature of billet, initial r   am speed and   bearing len gth, on m echanical pr operties of thixoextruded 7003 Al wrought   alloy were investigated. And, th   e study for thixoextrusion of 7003 Al wrought alloy was carried   out with respect to isotherm al holding tem perature and tim e during the partial remelting, especially in the low liquid fraction ($f_L<0.3$).

The liquid fraction and average grain size with   respect to isotherm al holding tim e were alm ost uniform. It is considered very useful for thixoext rusion in term s of process control such as billet temperature control and actual extrusion time.

The maximum extrusion pressure of the thixoextrusion was three times lower than that of the hot extrusion. The tensile and yield st rengthes of the thixoextruded bar before aging treatm ent were lower than those of the hot extruded bar, while the elongation value of the thixoextruded bar was higher than that of the hot extr   uded bar. However, the tensil e and yield strengthes of the thixoextruded bar af ter aging tr eatment were sim ilar to hot extrud ed bar. Therefore, their low tensile and yield strengthes could be improved through appropriate heat treatments.

References

[1] M.C. Flem ing, "Behavior of m etal alloys in the sem isolid state," *Metall. Trans.*, A22 (1991) 957-981.
[2] D.H. Kirkwood, "Semisolid Metal Processing," *Inter. Mater. Rev.*, 39(5) (1994) 173-189.
[3] G. Chiarm etta, "Thixoform ing of autom obile com ponents," (Paper presented   at the 4 [th] Processing of Semi-Solid Alloys and Composites, Sheffield, England, 19, June 1996), 204-207.
[4] K. Kiuchi and S. Sugiyam a, "Mashy-state Extrusion, Rolling and Forging," (Paper presented at the 3 [rd] Processing of Se mi-Solid Alloys and Composites, Tokyo, Japan, 13, June 1994), 245-257.
[5] K.P. Young and R. Fitze, "Sem   i-Solid Me tal Cast Alum inum Autom otive Com ponents," (Paper presented at th e 3 [rd] Processing of Se mi-Solid Alloys and Com posites, Tokyo, Japan, 13, June 1994), 155-177.

[6] P. Kapranos, R.C. Gibson, D.H. Kirkwood a nd C.M. Sellars, "Induction heating and partial melting of high m elting point thix oformable alloys," (Pap er presen ted at the 4 [th] Processing of Semi-Solid Alloys and Composites, Sheffield, England, 19, June 1996), 148-152.
[7] N. Akba s and A. Turkeli, "Form ation of non -dendritic structure in 7075 wrought alum inum alloy by SIMA process and effect of heat treatm ent," (Paper presented at th e 4 [th] Processing of Semi-Solid Alloys and Composites, Sheffield, England, 19, June 1996), 71-74.

211

Research on electromagnetic shielding property of aluminum foam

Haijun Yu, Guangchun Yao, Yihan Liu, Guangjun Yang

School of Materials and Metallurgy, Northeastern University. P.O. Box 117[#], Shenyang, Liaoning, 110004, P.R. China

Keywords: aluminum foam, electromagnetic shielding, relative density

Abstract

Through adjusting foaming temperature, foaming time, heat preservation time and vesicant addition amount and other technological parameters, Al-Si closed-cell aluminum foams of different densities were prepared by using molten body transitional foaming process in Northeastern University of China. Testing its electromagnetic shielding effectiveness using method of falan coaxial, the results show that the shielding effectiveness of material is affected obviously by the frequency of electromagnetic interference. With the interference frequency increasing from 10 MHz to 600MHz, shielding effectiveness of aluminum foam decreases gradually; and increases when frequency is added from 600 MHz to 1500 MHz. The influence of relative density on electromagnetic shielding effectiveness is not obvious.

Introduction

With development of modern science and technology, electronic and electric equipment has been widely applied in people's everyday life, national economy as well as national defense departments. Electronic and electric equipment not only increases in number and variety but also develops towards the direction of miniaturization, digitization and high speed. People are more and more dependent on electronic products: they are everywhere in working environment as well as living environment, which is brought by the development of science and technology. At the same time this also brings forth new problems [1, 2]: when electronic and electric equipment works, it will produce some electromagnetic energy either useful or useless, affecting other equipment, system and organism. As a result, electromagnetic environment becomes increasingly complex, resulting in "electromagnetic pollution".

Electromagnetic shielding is a way to prevent the spreading of high-frequency electromagnetic field in the space. When the interference frequency is above 10KHz, electromagnetic shielding is needed to prevent the interference of electromagnetic wave. When the shielding material is below 10KHz, static shielding or magnetic shielding can be utilized. Low resistance rate metal material (i.e. copper or aluminum) can be used as shielding material to generate strong inducing current thus engendering the vortex reverse to the interference electromagnetic field, as a result counteracting part of the original electromagnetic field and preventing the spreading of electromagnetic field, so this can achieve good shielding effectiveness [1]. Aluminum foam is a new kind of environmental friendly material in that its density smaller than that of water, anti-press, anti-impact and possessing fine electromagnetic shielding effectiveness. It can not only shield electromagnetic wave, decreasing the harm to human beings, but also adapt to the need for computer and other highly secret instrument in case the secrets are leaked [3, 4]. The author tests the electromagnetic shielding effectiveness of Al-Si closed–cell aluminum foam [5, 6] prepared by molten body transitional process, and researches the influence of interference frequency and the relative density of Al-Si closed–cell aluminum foam on its electromagnetic shielding effectiveness.

Experimental

First, industrial Al-Si alloy and metal calcium are used to confect Al-Si6.5%-Ca3% alloy; calcium is added to increase the viscosity of the melt. Second, after the alloy is melted, increase the temperature to 650□, and stir for 5 minutes. Third, add TiH_2 (47μm), and foam after mixing uniformly, then transfer and cool. Adjust technical parameters to prepare Al-Si closed-cell

aluminum foam of different relative densities (ρ_0 = 0.33, 0.28, 0.15, 0.12).

Utilizing saw bed, flattener and other equipment to machine aluminum foam into four samples (d=140mm). Figure 1 are the macroscopic and microcosmic photos of Al-Si closed–cell aluminum foam samples for electromagnetic shielding test. It shows that cell structure of test samples are uniform, basically in round shape; cell wall connect cells and there are small holes in the walls that are shared by cells.

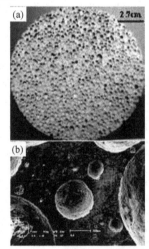

Figure 1. Testing sample: (a) macroscopic photograph; (b) microcosmic photograph

Electromagnetic shielding effectiveness of aluminum foam of different densities is tested by the method of falan coaxial at the 203 Institute of The Second Aviation Academy. The sketch map is shown in Figure 2.

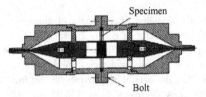

Figure 2. Schematic map of falan coaxial testing device

The samples are divided into benchmark samples and loading samples, the machining sketch map is Figure 3. The benchmark samples are divided into two parts. While testing, the circular part in the middle is set at the central conductor of the testing clamp, and the annular part is set at the exterior conductor falan of the testing clamp. The benchmark sample and the loading sample are of the same thickness, and for each sample the thickness difference between different points is less than 5% of its average thickness. The testing method is falan coaxial.

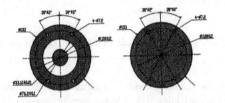

Figure 3. Dimensional drawing of benchmark sample and loading sample:
(a) basic specimen; (b) loading specimen

For a given source, dB is the ratio of electric field intensity or magnetic field intensity for a given point when there is no and is shielding.

$$SE = 20\lg\left(\frac{E_0}{E_1}\right) \tag{1}$$

$$SE = 20\lg\left(\frac{H_0}{H_1}\right) \tag{2}$$

In which SE is the shielding effectiveness (dB); E_0 or H_0 is the electric field intensity(V/m) or magnetic field intensity (A/m) when there is mo shield; E_1 or H_1 is the electric field intensity (V/m)or magnetic field intensity (A/m) when there is shield.

Results and Discussions

The Influence of Frequency on Shielding Effectiveness

Figure 4 shows the electromagnetic shielding effectiveness of aluminum foam, the relative density of specimen is 0.33, average diameter of cell is 2.4mm and wall thickness of cell is 0.6mm. The frequencies that people often meet are from 10MHz to 30000MHz; and taking the test precision into consideration, the chosen frequencies in this test are from 10MHz to 1500MHz. The figure shows that when the interference frequencies are from 10 to 600MHz, the electromagnetic shielding effectiveness decreases from 130dB to 40dB; while when the frequencies are from 600 to 1500MHz, the electromagnetic shielding effectiveness increases again.

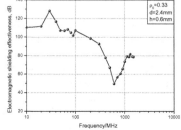

Figure 4. Influence of frequency on electromagnetic shielding effectiveness
of aluminum foam

The principle of electromagnetic shielding can be explained in combination with transmission line theory of electromagnetic shielding [1]. Electromagnetic wave is projected to Al-Si closed-cell aluminum foam whose thickness is t. When electromagnetic wave reaches the surface of aluminum foam, some is reflected by Al-Si conductor and the rest enters into the aluminum foam through one cell wall. After a distance of one cell diameter, it reaches another cell wall, some being reflected back to the incidence cell wall, some entering into adjacent cells through this one. Reflection and transmission continues until electromagnetic wave reaches the other side of aluminum foam board. The above mentioned process continues. So if the reflected electromagnetic wave energy when it just enters the aluminum foam board is called reflected wastage; the wastage of the transmission wave when it transmits in the board is called absorbed wastage; and the wastage that are caused by the repeated reflection within the aluminum foam board is called repeated reflective wastage [7,8], the electromagnetic shielding effectiveness is [1]:

$$SE=A+R+B \qquad (3)$$

In which, A is the absorbed wastage; R is the reflected wastage; B is the repeated reflective wastage. Usually when A>10dB, repeated reflective wastage can be omitted. The whole shielding effectiveness of aluminum foam boar is over 40dB, so the above formula can be

written as:
$$SE = A + R \tag{4}$$
$$A = 0.131 t \sqrt{f \mu_r \sigma_r} \tag{5}$$
$$R = 168 + 10 \lg \left(\frac{\sigma_r}{\mu_r f} \right) \tag{6}$$

t is the thickness of aluminum foam (mm); σ_r is the relative conductance of the shielding material to that of copper; μ_r is the relative magnetic conductance of the material; f is the frequency of the electromagnetic wave (Hz). Equations (5) and (6) indicate that absorbed wastage and reflected wastage are influenced not only by material itself but also by the frequency of the interference electromagnetic wave. For a given aluminum foam, shielding effectiveness mainly depends on the frequency of the interference electromagnetic wave. With the increase of the interference frequency, absorbed wastage increases while reflected wastage decreases. Combining Figure 4, it can be seen that when the frequency ranges from 10MHz to 600MHz, the decrease of reflected wastage is larger than the increase of absorbed wastage, so the integrated shielding effectiveness drops; when the frequency is from 600Mhz to 1500MHz, the decrease of reflected wastage is less than the increase of absorbed wastage, so the integrated shielding effectiveness rises.

The Influence of Relative Density on Shielding Effectiveness

As the wave length in this test is from 0.5m to 30m which belongs to long wave, the influence of the apertures which are only several millimeters in diameter on shielding effectiveness can be omitted. The electromagnetic shielding effectiveness of aluminum foam board can be judged qualitatively through the relative formula of the entity material [9]. Figure 5 is the electromagnetic shielding effectiveness of aluminum foam of three different relative densities under different magnetic field. The figure shows that the electromagnetic shielding effectiveness of aluminum foams of different densities does not change much and has the same trend. This also testifies the influence of frequency on the electromagnetic shielding effectiveness of aluminum foam. Increasing the relative density of aluminum foam, the diameter of cell in unit volume becomes small, the number of cells rises, and the area of the conductive section increases. According to ohm law, the conductance of aluminum foam remains the same if both conductive section and density increase; the wastage factor remains the same induced from equation (5), and the wave impedance also stays the same from equation (6). Equation (4) shows aluminum foam of different relative density has different electromagnetic shielding effectiveness.

Conclusions

(1) Al-Si closed-cell aluminum foam has good electromagnetic shielding effectiveness. When the interference frequency increases from 10MHz to 600MHz, absorbed wastage increases, and reflected wastage decreases, so as a result, the integrated shielding effectiveness drops. When the frequency further increases from 600MHz to 1500MHz, the integrated shielding effectiveness gains.

(2) Aluminum foam with different relative density keeps the same conductance because when density increases the conductive section also increases. Taking absorbed wastage and reflected wastage into consideration, the integrated shielding effectiveness also does not change. So relative density has little influence on the electromagnetic shielding effectiveness.

References

1. H.M. Lu, *Engineering EMC* (Xi'an: xi'an electron university of science and technology press, 2003), 2.
2. Y.F. Zhao et al., *Suppression technique of the electromagnetic radiation* (Beijing: China railroad press, 1990), 265.
3. A.G. Evans, J.W. Hutchinson and M.F. Ashby, "Multifunctionality of cellular metal systems," *Prog Mater Sci* , 43(1999), 171-221.
4. J. Banhart, "Manufacture, characterisation and application of cellular metals and metal foams," *Prog Mater Sci*, 46(2002), 559-632.
5. H.J. YU et al., "Influence of relative density on compressive behavior of Al-Si closed-cell aluminum foam," *J Northeastern University (Natural Science)*, 21(6)(2006), 1126-1129.
6. H.J. Yu, G.C. Yao and Liu Y H, "Research of tensile property of Al-Si closed-cell aluminum foam," *Trans Nonferrous Met Soc China, 2006*, 16(5)(in press).
7. Y. Feng, H.W. Zheng and Z.G. Zhu, "The microstructure and electrical conductivity of aluminum alloy foams," *Mater Chem Phys*, 78(2002), 176-201.
8. Y. Feng et al., "Electromagenetic shielding effectiveness of closed-cell aluminum alloy foams," *The Chinese Journal of Nonferrous Metals*, 14(2004), 1-4.
9. H.P. Degischer and B. Kriszt, *Handbook of Cellular Metals: Production, Processing, Applications* (Austria: Wiley-VCH press, 2002), 178.

AUTHOR INDEX

SUBJECT INDEX

T

V

W